Dynamic Offset Compensated CMOS Amplifiers

ANALOG CIRCUITS AND SIGNAL PROCESSING SERIES

Consulting Editor: **Mohammed Ismail.** ***Ohio State University***

For other titles published in this series, go to
www.springer.com/series/7381

Dynamic Offset Compensated CMOS Amplifiers

Johan F. Witte, Kofi A.A. Makinwa, Johan H. Huijsing
Delft University of Technology, the Netherlands

Springer
Boston/Dordrecht/London

Dr. Johan F. Witte
Delft University of Technology
Dept. Electrical Engineering
Mekelweg 4
2628 CD Delft
Netherlands
frerik.witte@nsc.com

Prof. Kofi A.A. Makinwa
Delft University of Technology
Dept. Electrical Engineering
Mekelweg 4
2628 CD Delft
Netherlands
k.a.a.makinwa@tudelft.nl

Prof. Johan H. Huijsing
Delft University of Technology
Dept. Electrical Engineering
Mekelweg 4
2628 CD Delft
Netherlands
j.h.huijsing@tudelft.nl

ISBN 978-90-481-2755-9 e-ISBN 978-90-481-2756-6
DOI 10.1007/978-90-481-2756-6
Springer Dordrecht Heidelberg London New York

Library of Congress Control Number: 2009926941

Printed on acid-free paper

Springer is part of Springer Science+Business Media (www.springer.com)

Preface ix

Acknowledgements xi

1. Introduction 1

1.1 Motivation 1
1.2 Offset 3
1.2.1 Drain current mismatch 4
1.2.2 Folded cascode amplifier offset 5
1.2.3 Minimizing offset 6
1.3 Challenges 7
1.4 Organisation of the book 8
1.5 References 10

2. Dynamic Offset Compensation Techniques 13

2.1 Introduction 13
2.2 Auto-zero amplifiers 14
2.2.1 Output offset storage 14
2.2.2 Input offset storage 16
2.2.3 Auxiliary amplifier 17
2.2.4 Noise in auto-zero amplifiers 19
2.3 Chopper amplifiers 23
2.3.1 Noise in chopper amplifiers 25
2.3.2 Chopped operational amplifier in a feedback network 26
2.3.3 Charge injection effects in chopper amplifiers 27
2.4 Chopped auto-zeroed amplifier 29
2.5 Switching non-idealities 31
2.5.1 Charge injection reduction tactics 33
2.5.2 Charge injection suppression circuits 36
2.6 Conclusions 40
2.7 References 40

3. Dynamic Offset Compensated Operational Amplifiers 43
3.1 Introduction 43
3.2 Ping-pong operational amplifier 44
3.3 Offset-stabilized amplifiers 45
3.3.1 Auto-zero offset-stabilized amplifiers 47
3.3.2 Chopper offset-stabilized amplifiers 48
3.3.3 Frequency compensation 50
3.3.4 Chopper stabilized amplifiers with ripple filters 55
3.3.5 Chopper and auto-zero stabilized amplifiers 58
3.4 Chopper offset-stabilized chopper amplifiers 59
3.4.1 Iterative offset-stabilization 61
3.5 Conclusions 63
3.6 References 64

4. Dynamic Offset Compensated Instrumentation Amplifiers 67
4.1 Introduction 67
4.1.1 Current-feedback instrumentation amplifiers 69
4.2 Dynamic offset compensated instrumentation amplifiers 74
4.2.1 Chopper instrumentation amplifier 75
4.2.2 Auto-zeroed instrumentation amplifier 76
4.2.3 Ping-pong instrumentation amplifier 78
4.2.4 Ping-pong-pang instrumentation amplifier 78
4.2.5 Offset-stabilized instrumentation amplifiers 79
4.2.6 Chopper offset-stabilized chopper instrumentation amplifier 82
4.3 Conclusions 82
4.4 References 82

5. **Realizations of Operational Amplifiers** 85
5.1 Introduction 85
5.2 Chopper offset-stabilized operational amplifier 86
5.2.1 Topology 86
5.2.2 Circuits 91
5.2.3 Measurement results 98
5.3 Chopper and auto-zero offset-stabilized operational amplifier 104
5.3.1 Topology 104
5.3.2 Circuits 107
5.3.3 Measurement results 112
5.4 Conclusions 115
5.5 References 116

6. **Realizations of Instrumentation Amplifiers** 117
6.1 Introduction 117
6.2 Low-offset indirect current-feedback instrumentation amplifier 118
6.2.1 Introduction 118
6.2.2 Topology 118
6.2.3 Circuits 122
6.2.4 Measurement results 124
6.3 High-side current-sense amplifier 129
6.3.1 Current-sensing 129
6.3.2 Topology 133
6.3.3 Circuits 138
6.3.4 Measurement results 143
6.4 Conclusions 147
6.5 References 149

7. **Conclusions and Future Directions** 151
7.1 Conclusions 151
7.2 Future directions 151
7.3 References 153

A. Layout Issues 155
A.1 Introduction 155
A.2 Chopper layout 157
A.3 Clock shielding 160
A.4 Conclusion 162
A.5 References 162

About the Authors 163

Index 167

Preface

CMOS amplifiers suffer from relatively poor offset specifications. Since the 1980s techniques have been explored to calibrate for this offset, or to let the amplifier itself compensate for its offset in some way or another. This latter approach is often done dynamically during operation of the amplifier, hence the name "dynamic offset compensation". This thesis describes the theory, design and realization of dynamic offset compensated CMOS amplifiers. It focuses on the design of general-purpose broadband operational amplifiers and instrumentation amplifiers.

Two distinguishable offset compensation techniques are described in chapter 2: auto-zeroing and chopping. Several topologies are discussed, in chapter 3 which can be used to design broadband dynamic offset-compensated operational amplifiers as well as instrumentation amplifiers, which are described in chapter 4. Four implementations are discussed in this book: two low-offset broadband operational amplifiers in chapter 5, and chapter 6 discusses a low-offset instrumentation amplifier, and a low-offset current-sense amplifier, which can sense battery currents at a 28V rail.

J.F. Witte
K.A.A. Makinwa
J.H. Huijsing
Delft, December 2008

Acknowledgements

This book started as a Ph.D. thesis written at the Electronic Instrumentation Laboratory of Delft university of technology, where I spent an productive, learningfull period of more than 6 years obtaining both my M.Sc. and Ph.D. degrees. I would start by thanking a lot of people, to whom I am indebted.

Firstly, I would like to thank my inspirators Han Huijsing and Kofi Makinwa. I am grateful to Han for introducing me into the field of precision amplifiers. I want to thank Kofi for giving me good advice and proofreading my publications.

Secondly, I would like to thank the people who in my opinion keep the university's wheels turning. Money makes the world go round and I would like to thank Willem van der Sluys for guiding every person of the laboratory through the financial bureaucracy. He even does it with a smile on his face. Without tools an engineer would only be a philosopher, and, therefore, I thank Antoon Frehe for keeping the computer servers in the air, despite failing and leaking air conditioners. My thanks also go to Evelyn, Ingeborg, Inge, Trudie, Pia, Helly and Joyce whose administrative support kept the group running through the first years of my M.Sc. and Ph.D. projects, and my thanks go to Ilse and Joyce who continue to keep the group running thanks to their ongoing administrative support.

Thirdly, I would really like to thank all the people who helped me during the design and measurements of my amplifiers. I want to thank Ger de Graaf, who has also defeated me quite often in our regular tennis matches. I want to thank Maureen Meekel, who even saw me crying once. Special thanks go to Piet, Jeff, Jeroen and Zu-Yao for helping me with various measurement problems. I also want to thank Harry Kerkvliet, who sadly enough passed away during my project, but he used to be a great help when a student needed equipment.

Special thanks also go to my former roommates Vladimir and Peter, and my fellow roommates Davina, Gayathri and Eduardo. Thanks also go to Michiel, Martijn, and Paulo with whom I have also enjoyed some vacations as well as tough technical discussions. I also have to thank the current group members Mahdi Kashmiri, Caspar van Vroonhoven, Rong Wu, and Andre Aita for many interesting discussions.

I would also like to thank all the people from Maxim semiconductor, who helped me with the implementation of the current-sense amplifier. I thank Paul and Bill for getting the project started, Matt Kolluri for helping me through my first real product design cycle, Jennifer for her layout efforts, Ray, Mike and Brian for their help in testing, and Rich for keeping the project going.

I also thank my former house-mate, Rob. I really thank him for maintaining a social circle. He taught me to drink whisky. We have brewed some mead and together with Martijn, Bas and Marc we slayed a dragon or two. Fun and friendship are necessary parts of life.

I also want to thank my family members. I especially want to thank my father for supporting me in my education. My aunt Corry for giving me advice over the years. I also would like to thank my mother. If you are able to raise a child to become an engineer, or even a doctor, then you really haven't been a bad mother after all.

Finally I want to thank my girlfriend Sophie with whom I have struggled through the last parts of this long and hard quest. Doing a Ph.D. is also a burden on your most loved ones. She has carried that burden.

J.F. Witte
Delft, December 2008

Introduction

1

1.1 Motivation

Low-offset amplifiers are needed in measurement systems. Typical applications include the read-out electronics of strain gauges, thermocouples, piezoelectric sensors, Hall sensors, or photo diodes. The signals generated by these devices are small, sometimes at the microvolt level. From an economical point of view, CMOS is the preferred technology for designing analog circuits, since it is relatively low cost and it enables the integration of low-power digital signal processing. This, in turn, makes the realization of complex mixed-signal systems feasible.

However, the input offset of typical CMOS amplifiers is at the millivolt level, limiting their accuracy severely. This compromises their usefulness in measurement systems. Therefore, techniques have been developed to solve this input offset problem. The need for precision electronics is the driving force behind a continuous effort to reduce the offset of CMOS amplifiers.

Calibration during production or trimming by the user would be the obvious solution to achieve a low offset, however, offset-trimmed CMOS amplifiers still suffer from offset drift over temperature and time. This offset drift will be an accuracy limit. Another method is to compensate for the offset dynamically, by implementing extra on-chip dynamic offset compensation circuitry in amplifiers. Because these techniques continue to compensate for the offset during the lifetime of the device, slow variations of the offset will also be compensated. Thus, offset drift over time and temperature will be strongly reduced.

Furthermore, considering the current trend towards lower supply voltages, offset in typical low-voltage CMOS amplifiers is becoming an increasingly more important limit in accuracy and dynamic range. Moreover, it can be predicted that knowledge about dynamic offset compensation techniques will become a necessity for future analog designers.

There are two different dynamic offset compensation techniques that can be distinguished, auto-zeroing and chopping [1.1]. Auto-zeroing is a sampling technique in which the offset is measured during one sampling phase and subtracted during another sampling phase. During the measurement phase, the amplifier cannot be used to amplify the input signal, which makes auto-zeroing difficult to implement in a continuous-time amplifier. Chopping, on the other hand, is a frequency modulation technique in which the signal and offset are modulated to different frequencies. In this way the offset can be distinguished from the signal, after which the offset is filtered out. This filter requirement makes it difficult to design a broadband amplifier.

The chopping technique was already explored in the late 1940s [1.2], when the signal of an amplifier implemented with vacuum tubes was modulated using mechanical switches. The auto-zero technique is probably much older. However, it was implemented in a monolithic amplifier in the early 1970s [1.3]. Both chopping and auto-zeroing techniques can be implemented in integrated circuits because of the availability of very good MOS switches. Dynamic offset compensated operational amplifiers became commercially available in the early 1980s [1.4] and implementations based on those early topologies are still available [1.5].

In recent years, many new developments have been made. For instance, a chopper offset-stabilized operational amplifier with a very good noise-power ratio has been developed [1.6] and commercialized [1.7], and a low-offset high-voltage device has been commercialized [1.8]. A more detailed overview of the many developments in this field will be presented in chapters 2, 3 and 4.

This book focuses on dynamic offset compensation techniques used in broadband CMOS amplifiers. In chapter 3 topologies are shown where auto-zero and chopping techniques are used in multi-path topologies [1.9]. In these topologies a low-frequency path is used to obtain a low offset, while a high-frequency path is used to obtain a high gain bandwidth product. This technique is called offset-stabilization, because, the offset of the high-frequency path is stabilized by the low-frequency path. The implementations described in chapters 5 and 6 focus on general-purpose feedback amplifiers with a gain bandwidth product of approximately 1 MHz and an offset in the μV range. Apart from operational amplifiers, indirect current-feedback instrumentation amplifiers [1.10] are also discussed. In contrast to traditional three-operational-amplifiers instrumentation amplifiers, such amplifiers isolate common-mode input and output voltages.

1.2 Offset

Before dynamic offset compensation techniques are discussed, it makes sense to discuss the nature and origins of offset in CMOS amplifiers. Input offset in a system is generally defined as the input level that forces the output level to go to zero. For an amplifier, as shown in figure 1-1, the input offset is the differential input voltage that forces the output voltage to go to zero.

Although offset is a DC parameter it can drift over time and temperature. This offset drift is usually specified in datasheets. The DC power supply rejection ratio (PSRR) and common-mode rejection ratio (CMRR) can be defined by:

$$PSRR = \frac{\Delta V_{DD}}{\Delta V_{os}} \text{ and } CMRR = \frac{\Delta V_{CM}}{\Delta V_{os}} . \quad (1\text{-}1)$$

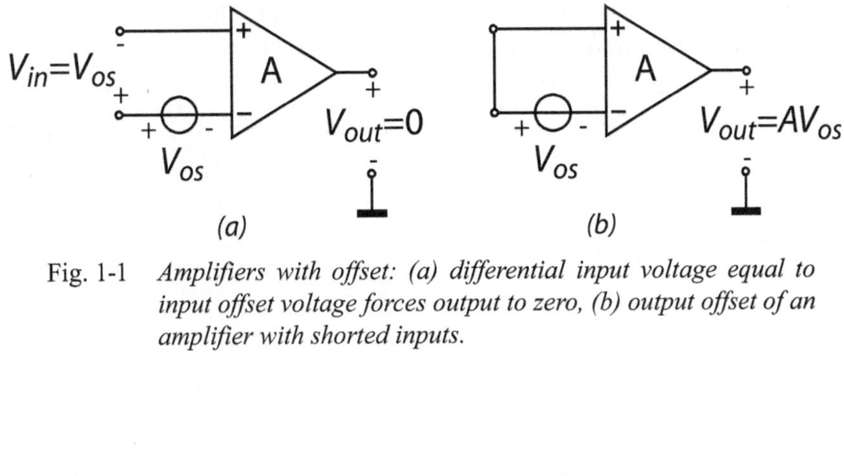

Fig. 1-1 *Amplifiers with offset: (a) differential input voltage equal to input offset voltage forces output to zero, (b) output offset of an amplifier with shorted inputs.*

Where ΔV_{DD}, ΔV_{os}, and ΔV_{CM} are the changes in power-supply voltage, input offset voltage, and input common-mode voltage, respectively. From these equations it can be seen that the offset can also change due to changing input common-mode and power supply voltages. A 100 dB CMRR means that the offset will shift 10 μV when the input common-mode changes by 1V.

It can be assumed that variations in the parameters of MOS transistors causes their drain current to vary, which ultimately causes input-referred offset voltage. In the following section, the input offset voltage of the commonly used folded-cascode operational amplifier is analysed. First, the mismatch dependency of the drain current will be derived.

1.2.1 Drain current mismatch

When operating in the strong inversion region, the drain current of a MOSFET can be described by:

$$I_D \approx \frac{1}{2}\mu C_{ox}\frac{W}{L}(V_{gs}-V_T)^2 = \beta(V_{gs}-V_T)^2 \,, \tag{1-2}$$

in which μ is the charge carrier mobility, C_{ox} is the normalized oxide capacitance, W is the channel width and L is the channel length of the MOS transistor, V_T is the threshold function, V_{gs} is the applied gate-source voltage and β is the transconductance factor. The variation in drain current caused by a threshold voltage mismatch will then be given by:

$$\frac{\delta I_D}{\delta V_T} = \frac{\delta I_D}{\delta V_{gs}} = -g_m \approx -\sqrt{2\mu C_{ox}\frac{W}{L}I_D} = -2\sqrt{\beta I_D} \approx \frac{2I_D}{V_{gs}-V_T} \,, \tag{1-3}$$

in which g_m is the transconductance of the transistor. The variation in drain current caused by a transconductance factor mismatch can be given by:

$$\frac{\delta I_D}{\delta \beta} \approx (V_{gs}-V_T)^2 \approx \frac{I_D}{\beta} \,. \tag{1-4}$$

When operating in the weak inversion region, the drain current of a MOSFET can be described by

$$I_D \approx I_s e^{\frac{V_{gs}-V_T}{nV_{th}}} = 2n\mu C_{ox}\frac{W}{L}V_{th}^2 e^{\frac{V_{gs}-V_T}{nV_{th}}} = 4n\beta V_{th}^2 e^{\frac{V_{gs}-V_T}{nV_{th}}}, \qquad (1\text{-}5)$$

in which I_s is the specific current, n is the weak inversion slope factor, and V_{th} is the thermal voltage, which is approximately 25 mV at room temperature. The implementations presented in this book were designed with 0.7 and 0.8 μm MOS processes. For these processes, n has a value of approximately 2. For more advanced submicron processes this value could approach 1.2. In the weak inversion region, the variation in drain current caused by a threshold voltage mismatch will then be given by:

$$\frac{\delta I_D}{\delta V_T} = -\frac{\delta I_D}{\delta V_{gs}} = -g_m \approx -\frac{I_D}{nV_{th}}. \qquad (1\text{-}6)$$

In weak inversion, the variation in drain current caused by a transconductance factor mismatch will then be given by:

$$\frac{\delta I_D}{\delta \beta} \approx 4nV_{th}^2 e^{\frac{V_{gs}-V_T}{nV_{th}}} \approx \frac{I_D}{\beta}. \qquad (1\text{-}7)$$

From equations (1-4) and (1-7) it can be concluded that the effect of the transconductance factor mismatch is proportional to the drain current in both weak and strong inversion. Similarly, the effect of threshold voltage mismatch is proportional to the transconductance of the transistor in both weak and strong inversion.

1.2.2 Folded cascode amplifier offset

In figure 1-2, a folded cascode amplifier is shown. It can be assumed that the cascode transistors M_7, M_8, M_9, and M_{10} do not contribute to the offset.

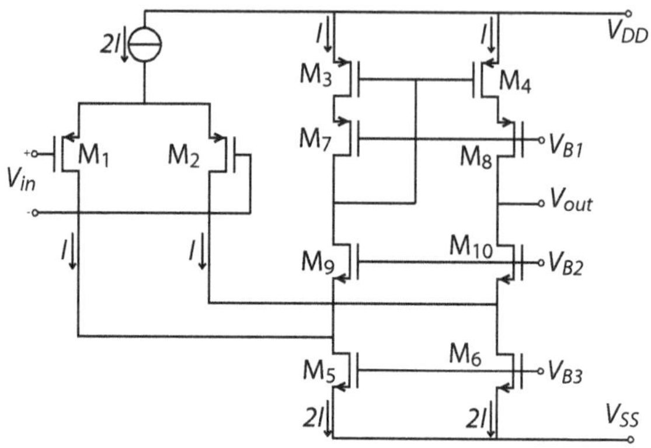

Fig. 1-2 *Folded cascode operational amplifier.*

When the effects of the transconductance factor mismatch and threshold voltage mismatch of the three transistor pairs M_{1-2}, M_{3-4}, and M_{5-6} are superposed, the offset can be expressed as:

$$V_{OS} = \Delta V_{T1,2} + \frac{g_{m3,4}}{g_{m1,2}}\Delta V_{T3,4} + \frac{g_{m5,6}}{g_{m1,2}}\Delta V_{T5,6} + \frac{I}{g_{m1,2}}\left(\frac{\Delta\beta_{1,2}}{\beta_{1,2}} + \frac{\Delta\beta_{3,4}}{\beta_{3,4}} + 2\frac{\Delta\beta_{5,6}}{\beta_{5,6}}\right), \quad (1\text{-}8)$$

where ΔV_T and $\Delta\beta$ are the differences in threshold voltages and transconductance factors of the indicated transistors respectively. The offset can be minimized by reducing the transconductance of the current sources M_5 and M_6 and of current mirror M_3 and M_4, meaning that they should work in strong inversion. To obtain an optimal offset the input stage transistors should be given a large transconductance and their ratio I_D/g_m should be as small as possible, meaning that the input transistors M_{1-2} should work in weak inversion, which indicates that $I_D/g_m = V_{th}n$, which is typically 50 mV at room temperature.

1.2.3 Minimizing offset

Variations in threshold voltages and transconductance factors are caused by mismatch. This is defined as the process of time-independent random variations in physical quantities of identical designed devices [1.11]. Moreover, it is assumed that the transconductance factors β and the threshold voltage V_T have a stochastic variation due to mismatch. The standard deviation of the threshold voltage may be approximated by:

$$\sigma^2(V_T) = \frac{A_{V_T}^2}{WL} + S_{V_T}^2 D^2 \ , \tag{1-9}$$

where $A_{V_{th}}$ and $S_{V_{th}}$ are process-related constants and D is the distance between two transistors [1.11]. Therefore, it can be seen that threshold variations are inversely proportional to the square root of the transistor area and proportional to the distance between transistors. The relative standard deviation of the transconductance factor can be written as

$$\frac{\sigma^2(\beta)}{\beta^2} = \frac{A_W^2}{W^2 L} + \frac{A_L^2}{WL^2} + \frac{A_{C_{ox}}^2}{WL} + \frac{A_\mu^2}{WL} + S_\beta^2 D^2 \approx \frac{A_\beta^2}{WL} + S_\beta^2 D^2 , \tag{1-10}$$

where A_W, A_L, $A_{C_{ox}}$, A_μ and S_β are process related constants and $A_\beta^2 = A_{C_{ox}}^2 + A_\mu^2$ [1.11]. In many processes only A_β is specified, but this is not sufficient to estimate the mismatch of transistors with small W or L in which A_W and A_L are more dominant mismatch sources. The offset of a CMOS folded cascode gain stage as shown in figure 1-2 is typically 5–10 mV.

If, for all transistors in a folded cascode amplifier, $\Delta V_T < 0.5\,\text{mV}$, $\Delta\beta/\beta < 0.15\%$, $g_{m1,2}/I = 20\,\text{V}^{-1}$ and $g_{m1,2} = 5g_{m5,6} = 10g_{m3,4}$, then the obtained offset would be $V_{OS} < 0.95\,\text{mV}$. In order to obtain $\sigma(V_T) < 0.125\,\text{mV}$, which is needed to reach a 4σ value smaller than 0.5 mV, in a 0.7 μm process where $A_{V_T} = 10\,\text{mV}/\mu\text{m}$, transistors with an effective area of 6400 μm^2 are needed. These can be regarded as very large transistors.

Instead of increasing the transistor size to improve offset behaviour, it can be considered to add extra circuitry for offset trimming or dynamic offset compensation.

1.3 Challenges

The main topic of this book is to design dynamic offset compensated CMOS amplifiers, with an approximately 1 MHz gain bandwidth product and an offset in the μV range. This was already done in [1.4], where an auto-zero technique was used in a low-frequency path to stabilize the offset of a high-frequency path. In this book this technique will be called auto-zero offset stabilization. It is known that chopper amplifiers have a superior noise performance but a limited bandwidth because of the filter requirement [1.1] [1.12]. The main scientific challenge, addressed in this book, is to use the chopper technique in a low-frequency path to stabilize the offset of a high-frequency path [1.6] [1.13] [1.14], leading to a superior noise specification. This technique will be called chopper offset stabilization in this book.

There are several challenges associated with the design of dynamic offset compensated amplifiers. The added offset compensation circuitry probably increases the power consumption, which customers do not like. Secondly, extra circuitry means extra design time, which design managers do not like. Although analog designers are generally intelligent beings, their intellectual flexibility is still limited, and will stay limited. Thus, the more complex the topology, the more mistakes a designer can make, and the longer a design cycle takes, or the lower the yield. The die size is somewhat related to the complexity of a topology. However, in practice, capacitors and resistors occupy large chip areas. Therefore, the number of capacitors and resistors used in a design should be limited. To keep power consumption within limits, a topology should not have too many current consuming blocks. Therefore, in this work an attempt has been made to design amplifiers with a reasonable die size, power consumption and complexity.

An additional challenge, addressed in section 6.3, is to obtain a low-offset voltage for a high-voltage device. In 2003, 5.5 V supply voltage low-offset wide-bandwidth amplifiers were available [1.5][1.15][1.16]. However, a high-voltage low-offset part was not. Although since 2007, a low-offset current-sense amplifier has been commercially available [1.8].

In conclusion, the design of a dynamic offset compensated amplifier should be relatively simple, while the implementation should use a relatively small silicon area and requires low-power consumption.

1.4 Organisation of the book

This work has been divided into seven chapters and one appendix. Following this introduction, chapter 2 describes the two dynamic offset compensation techniques, auto-zeroing and chopping. These two techniques can be considered as the only dynamic offset compensation techniques. However, they have a limited usability in broadband general-purpose amplifiers. Therefore, chapter 3 goes a step further and describes how these techniques can be used in multi-path operational amplifiers [1.9] creating broadband dynamic offset compensated operational amplifiers. This theory will be extended to current-feedback instrumentation amplifiers in chapter 4. Two chapters are devoted to the implementation of broadband dynamic offset compensated CMOS amplifiers. Chapter 5 describes two designs of operational amplifiers, and in chapter 6 two instrumentation amplifier topologies are described. The book ends with conclusions and some recommendation for future research.

The first implementation described in chapter 5 is an operational amplifier designed as a feasibility study for the chopper offset-stabilization technique. It operates with an external clock. This amplifier obtained a submicrovolt offset with external clock frequencies of 4 kHz.

The second operational amplifier discussed in chapter 6 has an on-board oscillator and draws 300 μA of current, while obtaining a 30 nV/√Hz noise and a 1 MHz GBW with a 100 pF load capacitance. Both the auto-zero and chopper technique are used to obtain a low offset. The first instrumentation amplifier implementation in chapter 6 is actually the same operational amplifier with an extra input stage to support an indirect current-feedback instrumentation amplifier topology. It draws 340 μA of current while obtaining a 40 nV/√Hz noise and a 1 MHz GBW with a 100 pF load capacitance.

The last implementation discussed in this book is of a current-sense amplifier [1.17]. Both the auto-zero and chopper technique are used to obtain an offset of less than 5 μV. It has an input common-mode voltage range from 1.9 to 30 V and a DC common-mode rejection ratio of 143 dB.

In the process of designing these types of amplifiers, the author experienced critical problems with the layout of the amplifiers. In appendix A some practical layout issues are discussed, which could help future designs to be more successful.

1.5 References

[1.1] C.C. Enz, G.C. Temes, "Circuit techniques for reducing the effects of op-amp imperfections: autozeroing, correlated double sampling, and chopper stabilization", *Proceedings of the IEEE*, pp. 1584–1614, Nov. 1996.

[1.2] A.J. Williams, R.E. Tarpley, W.R. Clarck, "D-C amplifier stabilized for zero and gain", *Trans. AIEE*, Vol. 67, pp. 47–57, 1948.

[1.3] R. Poujois, B. Baylac, D. Barbier, J. Ittel, "Low-level MOS transistor amplifier using storage techniques", *IEEE ISSCC*, pp. 152–153, Feb. 1973.

[1.4] M.C.W. Coln, "Chopper stabilization of MOS operational amplifiers using feed-forward techniques", *IEEE JSSC*, Vol. 16, pp. 745–748, Dec. 1981.

[1.5] Intersil, "ICL7650, 2MHz, super chopper-stabilized operational amplifier", Datasheet, http://www.intersil.com Mar. 2008.

[1.6] R. Burt, J. Zhang, "A micropower chopper-stabilized operational amplifier using a SC notch filter with synchronous integration inside the continuous-time signal path", *IEEE JSSC*, pp. 2729–2736, Dec. 2006.

[1.7] OPA333 "1.8V, micropower CMOS operational amplifiers zero-drift series", Datasheet, www.ti.com, May 2007.

[1.8] Linear Technology, "LTC6102", Datasheet http://www.linear.com, July 2007.

[1.9] R.G.H. Eschauzier, L.P.T. Kerklaan, J.H. Huijsing, "A 100-MHz 100-dB operational amplifier with multipath nested Miller compensation structure", *IEEE JSSC*, pp. 1709–1717, Dec. 1992.

[1.10] B.J. van den Dool, J.H. Huijsing, "Indirect current feedback instrumentation amplifier with a common-mode input range that includes the negative roll", *IEEE JSSC*, pp. 743–749, July 1993.

[1.11] M.J.M. Pelgrom, A.C.J. Duinmaijer, A.P.G. Welbers, "Matching properties of MOS transistors", *IEEE JSSC*, pp. 1433–1439, Oct. 1989.

[1.12] A. Bakker, J.H. Huijsing, "*High-accuracy CMOS smart temperature sensors*", Dordrecht: Kluwer, 2000.

[1.13] J.F. Witte, K.A.A. Makinwa, J.H. Huijsing, "A CMOS chopper offset-stabilized opamp", *ESSCIRC*, pp. 360–363, Sep. 2006.

[1.14] J.F. Witte, K.A.A. Makinwa, J.H. Huijsing, "A CMOS chopper offset-stabilized opamp", *IEEE JSSC*, pp. 1529–1535, July 2007.
[1.15] Analog Devices, "AD8628", Datasheet, http://www.analog.com, Mar. 2008.
[1.16] Maxim, "MAX4238", Datasheet http://www.maxim-ic.com, Mar. 2008.
[1.17] J.F. Witte, J.H. Huijsing, K.A.A. Makinwa, "A current-feedback instrumentation amplifier with 5μv offset for bidirectional high-side current-sensing", *IEEE ISSCC*, pp. 74–75, Feb. 2008.

Dynamic Offset Compensation Techniques

2

2.1 Introduction

In this chapter, the theory underlying two dynamic offset compensation techniques, chopping and auto-zeroing [2.1] will be discussed. These techniques can be used in the design of both general purpose amplifiers and dedicated read-out electronics for sensors. By using these techniques the offset is compensated continuously. Therefore, these also help to reduce offset drift over temperature or time. Since low-frequency noise and DC offset can not be distinguished from each other by the dynamic offset compensation techniques, both techniques also have an effect on low-frequency noise.

A good rule of thumb is that a dynamic offset compensation technique can be expected to reduce the initial offset of an amplifier by a factor of 100 to 1000. When a circuit has a very strict offset specification, a combination of dynamic offset compensation techniques can be used.

Over the years, some confusion has arisen with regard to the naming of the different dynamic offset compensation techniques. In this book two techniques are distinguished, namely: auto-zeroing and chopping. These two techniques are described in this chapter. In chapter 3, offset stabilization will be introduced. In this technique the offset of a main amplifier is measured with a low-offset stabilization amplifier, which generates a correction signal to compensate for the offset of the main amplifier.

In this chapter, auto-zeroing is discussed first. This technique can also be described as time-domain modulation of offset, since the offset is measured at one moment in time, and at another moment the signal is measured, and the offset is subtracted from the signal. Afterwards, chopping is discussed. This technique can also be described as frequency-domain modulation of offset. The input offset and input signal are modulated to different frequencies after which the modulated offset can be filtered out.

2.2 Auto-zero amplifiers

During auto-zeroing, two phases in time can be distinguished: one auto-zero phase in which the offset of a system is measured and stored, and one signal phase in which the signal is amplified and the offset is subtracted from the signal.

2.2.1 Output offset storage

Probably the simplest way to implement dynamic offset compensation is to place a capacitor at the output of an amplification stage, as shown in figure 2-1. The capacitor C_1 is used to store the output referred offset and subtract it from the signal. Therefore, this technique is called auto-zeroing with output offset storage [2.2]. In the literature this technique is also known as open-loop offset cancellation [2.1]. During the sampling phase F_2, S_1 and S_4 are open and S_2 and S_3 are closed. During the signal phase F_1, S_1 and S_4 are closed and S_2 and S_3 are open.

During the sampling phase F_2 the voltage over the auto-zero capacitor C_1 can be expressed as:

$$V_{out1} = V_c = \mathrm{A} V_{os}, \qquad (2\text{-}1)$$

where A is the voltage gain of the amplifier, while during the signal phase F_1 the output voltage can be expressed as:

$$V_{out1} = (V_{in} + V_{os})\mathrm{A} \text{ and } V_{out2} = V_{out1} - V_c = \mathrm{A} V_{in}. \qquad (2\text{-}2)$$

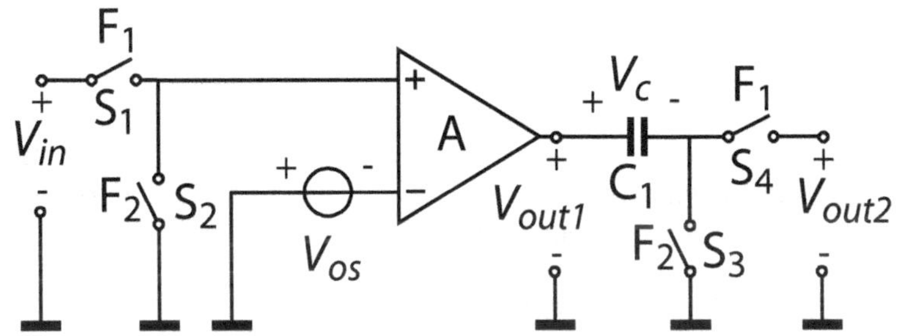

Fig. 2-1 *Auto-zeroed amplifier with output offset storage.*

This means that theoretically all offset is cancelled using this method. However, when the switches are implemented with MOS transistors they will inject charge into the auto-zero capacitance C_1. This charge injection will be described in more detail in section 2.5. However, it can be assumed that a charge q_{inj} is produced by the switch each time it closes, and an equal negative charge $-q_{inj}$ each time it opens.

At the end of the auto-zero phase, switch S_3 opens and switch S_4 closes. Therefore, a charge injection mismatch $q_{inj3}-q_{inj4}$ is fed into the storage capacitor. Thus:

$$V_{out1} = V_c = AV_{os} + \frac{(q_{inj3} - q_{inj4})}{C_1}. \qquad (2\text{-}3)$$

This results in a residual offset of:

$$V_{os,res} = \frac{(q_{inj3} - q_{inj4})}{AC_1}. \qquad (2\text{-}4)$$

The influence of this mismatch on the residual offset can be reduced by increasing the size of the capacitor. Since the auto-zero capacitor is at the output of the amplifier, these effects can be divided by the voltage gain when referred to the input. Leakage from capacitor C_1 during the amplification phase also causes residual offset. This effect can also be divided by the voltage gain.

A disadvantage of auto-zeroing with output offset storage is that the output range is reduced by $2AV_{os,max}$, where $V_{os,max}$ is the maximal input offset that can be expected of an uncompensated amplifier. When the voltage gain is 40 dB and $V_{os,max} = 10\,\mathrm{mV}$, then the output range is reduced by 2 V. Therefore, this technique can only be used in low-gain amplifiers, and not in high-gain general purpose amplifiers, but it can be useful in custom read-out electronics. However, a fully integrated amplifier with three cascaded auto-zeroed amplifiers with output offset storage has been reported in [2.1], these stages were later chopped as well [2.3].

2.2.2 Input offset storage

There is another auto-zeroing technique that is implemented with input offset storage [2.2]. This technique is also known as closed loop offset cancellation in the literature [2.1]. An implementation is shown in figure 2-2. During the sampling phase F_2, S_1 and S_4 are open while S_2 and S_3 are closed. During the signal phase F_1, S_1 and S_4 are closed while S_2 and S_3 are open.

During the sampling phase F_2, the voltage over the auto-zero capacitor C_1 can be expressed as:

$$V_c = \frac{A}{A+1} V_{os}, \tag{2-5}$$

while during the signal phase F_1 the output voltage can be expressed as:

$$V_{out} = (V_{in} + V_{os} - V_c)A = \left(V_{in} + \frac{1}{A+1} V_{os}\right)A. \tag{2-6}$$

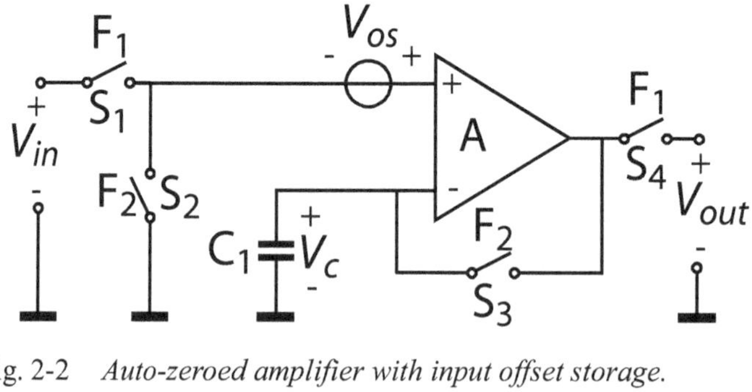

Fig. 2-2 *Auto-zeroed amplifier with input offset storage.*

Apart from the finite gain, charge injection is a source of residual offset. The channel charge of switch S_3 will again cause a voltage step across the auto-zero capacitor C_1. Considering the same analysis provided in the previous section, this results in a residual offset which can be expressed as:

$$V_{os,res} = \frac{V_{os}}{A+1} + \frac{q_{inj3}}{C_1}. \qquad (2\text{-}7)$$

The influence of S_3 on the residual offset can be reduced by increasing the size of the capacitor. In addition, the leakage of the capacitor C_1 during the amplification phase can cause residual offset. These two effects cannot be divided by the voltage gain because the auto-zero capacitor is already at the input. In differential circuits, the residual offset due to charge injection will also be reduced as will be explained in section 2.5.1.

This technique can be used to auto-zero a high gain operational amplifier. The residual offset is then dominated by the charge injection.

2.2.3 Auxiliary amplifier

Another technique for auto-zeroing which is less sensitive to charge injection is depicted in figure 2-3 [2.4]. In many cases, an amplifier will consist of a transconductance G_1 with an output impedance R_{out}. This transconductance G_1 has an offset voltage V_1. Phase F_1 is the signal phase in which the input signal is applied to the amplifier and the output signal is useful. Phase F_2 shorts the inputs of G_1 and its offset V_{os} causes an output current I_1 to flow. This output current is integrated on capacitor C_1. This capacitor drives an auxiliary input transconductance G_2, which causes an offset compensating

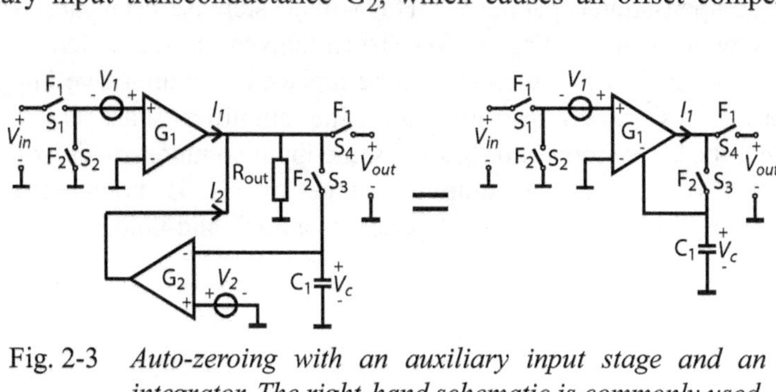

Fig. 2-3 *Auto-zeroing with an auxiliary input stage and an integrator. The right-hand schematic is commonly used.*

current I_2 to flow. The loop gain limits the offset reduction. In steady state during auto-zero phase F_2 the following equation applies:

$$V_C = V_1 G_1 R_{out} + (V_2 - V_C) G_2 R_{out} \quad \text{thus} \quad V_C = \frac{V_1 G_1 R_{out} + V_2 G_2 R_{out}}{1 + G_2 R_{out}}, \tag{2-8}$$

while during the signal phase F_1 the following equation applies:

$$V_{out} = (V_1 + V_{in}) G_1 R_{out} - V_C G_2 R_{out} = G_1 R_{out} V_{in} + \frac{V_1 G_1 R_{out}}{1 + G_2 R_{out}} + \frac{V_2 G_2 R_{out}}{1 + G_2 R_{out}}. \tag{2-9}$$

The residual offset due to finite gain can be expressed by:

$$V_{os,res,gain} = \frac{V_1}{1 + G_2 R_{out}} + \frac{1}{G_1 R_{out}} \frac{V_2 G_2 R_{out}}{1 + G_2 R_{out}} \approx \frac{V_1}{A_2} + \frac{V_2}{A_1}, \tag{2-10}$$

where A_2 and V_2 are the DC voltage gain and offset of G_2, which acts as an auxiliary input of G_1. Except for the limited voltage gain, additional residual offset is also caused by the charge injection of switch S_3 and the leakage of the capacitor C_1 during the signal phase. The charge injection causes the residual offset to become:

$$V_{os,res,inj} = \frac{G_2 q_{inj3}}{G_1 C_1}. \tag{2-11}$$

The output of G_1 has to switch between the output voltage and the voltage V_C over the compensation capacitor C_1. This voltage step itself can cause settling problems, which cause voltage spikes. To circumvent these problems another topology can be used in which C_1 can be replaced with an active integrator. This concept is shown in figure 2-4. The amplifier stage G_4 used to implement the active integrator has the same input common-mode voltage as the output stage G_3. The output voltage of G_1 is bounded at this common-mode voltage. Capacitor C_{2b} acts as a track-and-hold.

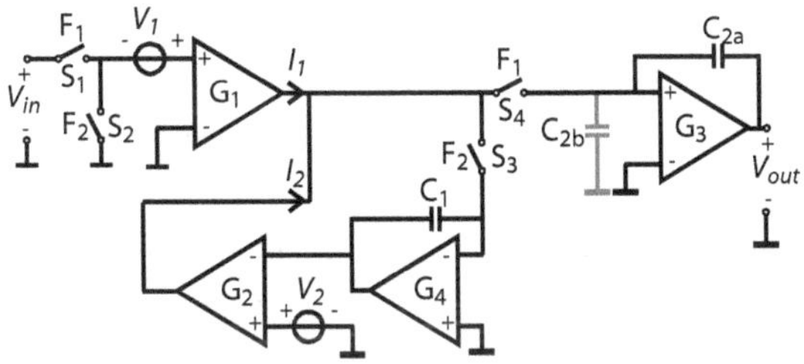

Fig. 2-4 *Auto-zeroing with an auxiliary input stage and an active integrator to circumvent voltage steps at the output of G_1.*

Because of the sampling action, auto-zeroing is a technique which is not suited for continuous time applications. Auto-zeroing itself is mainly used in switched capacitor circuits, which are already sampled systems [2.1]. Also when continuous-time operation is required, a ping-pong architecture can be used, in which two auto-zeroed amplifiers run parallel to each other, one being auto-zeroed and one being used to amplify the signal. This architecture will be discussed in section 3.2.

2.2.4 Noise in auto-zero amplifiers

The auto-zero technique cannot distinguish low-frequency noise from offset. Therefore, the noise behaviour over frequency of an auto-zeroed amplifier is also affected. A quantitative calculation of this effect has been provided by Enz [2.1] [2.5].

Intuitively, it can be assumed that when the amplifier depicted in figure 2-2 is auto-zeroed, the broadband noise of the amplifier is projected onto the capacitor C_1. At the end of the sampling phase the input offset and noise voltage are held on the capacitor, which means that all components of noise above the auto-zeroing frequency will fold back due to aliasing. As a result, the white noise in this band folds back to below the auto-zero frequency.

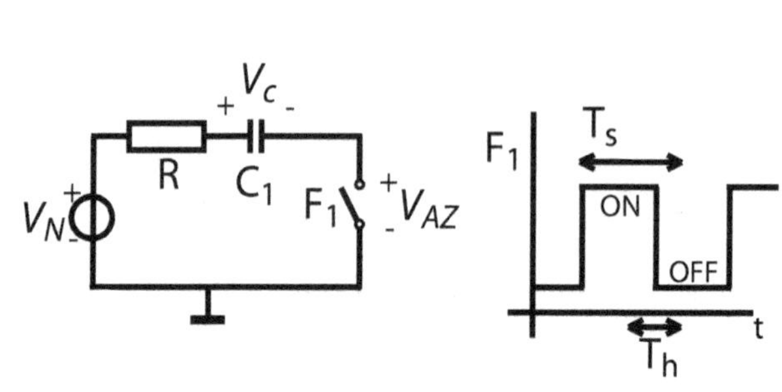

Fig. 2-5 *Equivalent auto-zero circuit.*

According to [2.1] an auto-zero amplifier, as depicted in figure 2-2, can be seen as the circuit, shown in figure 2-5. In this circuit, V_N represents the noise at the output of the amplifier in the auto-zero phase. It has been shown in [2.1] and [2.5] that the noise power spectral density (PSD) for the auto-zero voltage across the switch can expressed as:

$$S_{AZ}(f) = \sum_{n=-\infty}^{\infty} |H_n(f)|^2 S_N\left(f - \frac{n}{T_s}\right), \tag{2-12}$$

where $|H_n(f)|^2$ are transfer functions which model the folding effect of each band n, $S_N(f)$ is the noise PSD of V_N, and T_s is the sampling period. In other words infinite noise bands are folded on top of each other to form a new noise characteristic. Luckily the bandwidth is limited, otherwise the PSD would be unlimited. The transfer functions can be expressed by [2.1]:

$$|H_0(f)|^2 = d^2\left\{\left[1 - \frac{\sin(2\pi f T_h)}{2\pi f T_h}\right]^2 + \left[\frac{1-\cos(2\pi f T_h)}{2\pi f T_h}\right]^2\right\} \quad \frac{T_h}{T_s} = d, \tag{2-13}$$

where d is the duty cycle of hold time in the auto-zero action and for non-zero n

$$|H_n(f)|^2 = d^2\left\{\left[\frac{\sin(2\pi dn)}{2\pi dn} - \frac{\sin(2\pi f T_h)}{2\pi f T_h}\right]^2 + \left[\frac{1-\cos(2\pi dn)}{2\pi dn} - \frac{1-\cos(2\pi f T_h)}{2\pi f T_h}\right]^2\right\}, \tag{2-14}$$

where T_h is the hold time during which there is no sampling action. With these equations it has been proven [2.1] [2.5] that auto-zeroing cancels the

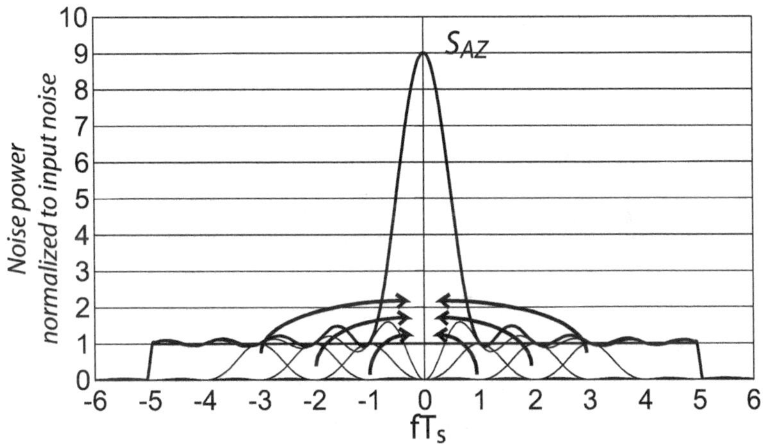

Fig. 2-6 *Sketched noise folding in auto-zeroing d=1.*

offset, or the DC component in the noise, and that the low-frequency noise of the resulting auto-zero amplifier is caused by the noise folding. It has also been shown that $1/f$ noise is effectively compensated for when the $1/f$ corner frequency is lower than the auto-zero frequency.

Since this looks rather complicated, however, it is advisable to spend a day or two working with a mathematical program like Maple, Matlab or Excel, to get a feeling for noise folding before starting a design for an auto-zero amplifier. Consider, for instance, an amplifier with an equivalent noise bandwidth equal to five times the auto-zero frequency. This has been sketched in figure 2-6, where $d=1$, i.e. for a sampling time of zero seconds. The side-bands fold back to the DC frequency. In this diagram it can be seen that the noise is folded nine times, one time for the baseband and four times for each side-band.

In figure 2-7a the transfer functions $|H_n(f)|^2$ are sketched for $d=1$. In figure 2-7b the $|H_n(f)|^2$ are sketched for $d=0.5$. It can be seen that the energy is more spread out over frequency. In figure 2-8 the noise folding has been sketched for $d=0.5$. The PSD amplitude around DC has been reduced by approximately d^2, and it can be seen that the PSD hits half the white noise level at $2fT_S$. The noise reduction can be explained by the noise power spreading out over a wider bandwidth. It is essential to note that this is the simulated noise PSD over the capacitor C_1 in figure 2-2, and that the input

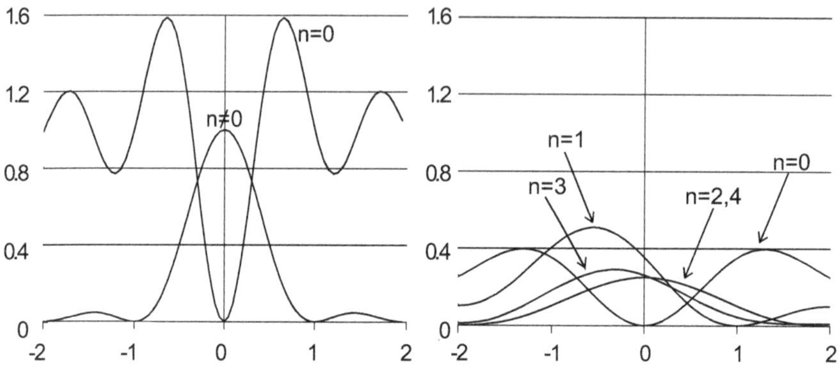

(a) $d = 1$ (b) $d = 0.5$

Fig. 2-7 $|H_0(f)|^2$ *sketched for* $n=0,1,2,3,4$ *with (a)* $d=1$ *and (b)* $d=0.5$.

signal power seen by the amplifier is also multiplied by the duty cycle *d*. Therefore, the equivalent PSD seen by the input signal at low frequencies can be expressed as:

$$S_{eq,AZ}\left(f<\frac{1}{T_s}\right)=\frac{S_{AZ}(f)}{d}\ . \tag{2-15}$$

This does not mean that the input noise power is reduced by increasing *d*. It only means that the noise PSD should be reduced for low frequencies. From figure 2-8, it can be concluded, that around two times the auto-zero frequency the input referred PSD of auto-zero amplifiers with a 50% duty cycle is equal to the white noise level of the amplifier.

During the auto-zero phase, when the amplifier is in a unity-gain setting, the equivalent white noise bandwidth is $0.5\pi g_m/C_1$, where g_m is the transconductance of the amplifier and C_1 is the auto-zero capacitor. When all equivalent white noise is folded back to the base-band, then it can be assumed from the analysis above that the auto-zero amplifier for low frequencies, i.e. $f<0.5/T_s$, has a folded white noise floor of:

$$V_{eqn,LF}^2\approx d\frac{\pi}{2\pi}\frac{g_m}{2C_1}T_sV_{n,th}^2=\frac{dT_sg_m}{4C_1}V_{n,th}^2, \tag{2-16}$$

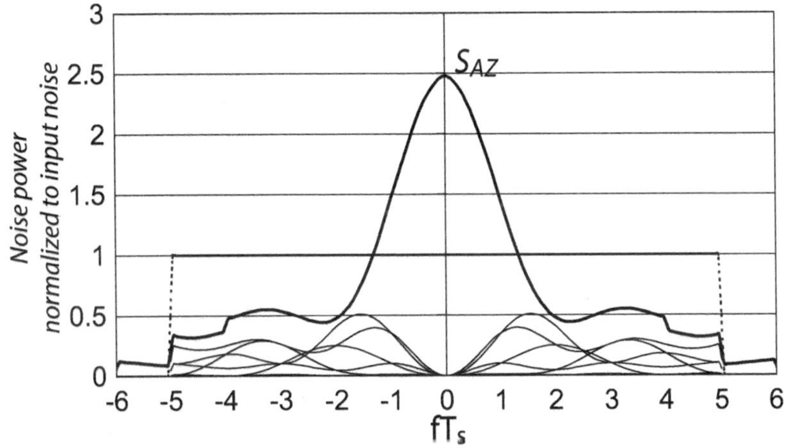

Fig. 2-8 *Sketched noise folding in auto-zeroing d=0.5.*

This means that the white noise folding can severely harm the low-frequency characteristics. For instance, if a transconductance of 200 μA/V is auto-zeroed on a 10 pF capacitor with a duty-cycle of 50%, then the –3dB bandwidth is 3.18 MHz, and the equivalent white noise bandwidth is 5 MHz. If the auto-zero frequency is 10 kHz, the noise at low frequencies would be a factor 250 larger than the white noise power, and for the noise signal voltage this would be a factor $\sqrt{250} \approx 16$. In the next section it will be shown that chopped amplifiers do not have this drawback.

2.3 Chopper amplifiers

While in auto-zero amplifiers the offset and input signal are time-modulated, in chopper amplifiers they are frequency-modulated. In chopper amplifiers the signal of interest and the offset signal are shifted to different frequencies.

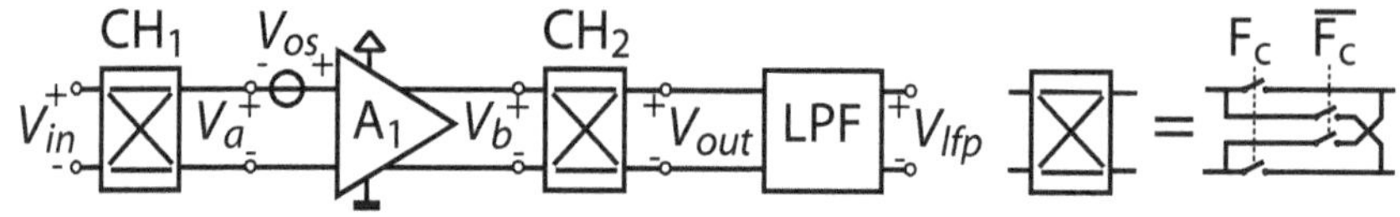

Fig. 2-9 *Chopper amplifier.*

In figure 2-9 a chopper amplifier is shown [2.1] [2.6]. This amplifier consists of a frequency modulator, or chopper CH_1, a voltage amplifier A_1, another chopper CH_2, and a low-pass filter (LPF). The chopper symbol is also depicted, which is a polarity switch driven by a square wave with a chopper frequency F_C.

The idealised waveforms are depicted in the time domain in figure 2-10. The first chopper is used to modulate the input signal to a higher frequency. Then the amplifier amplifies the modulated signal superposed on its own input noise sources. Lastly, the second chopper demodulates the amplified input signal, and also modulates the output noise and offset of the amplifier A_1.

The limited bandwidth of the amplifier A_1 is a fundamental cause of switching glitches, because the input signal is modulated almost ideally, while the amplified signal V_b is slightly delayed when it is demodulated, causing high frequency glitches. The combination of non-zero offset and the chopper action also give rise to a chopper ripple, which has the same frequency as the chopper clock and is proportional to the offset of the

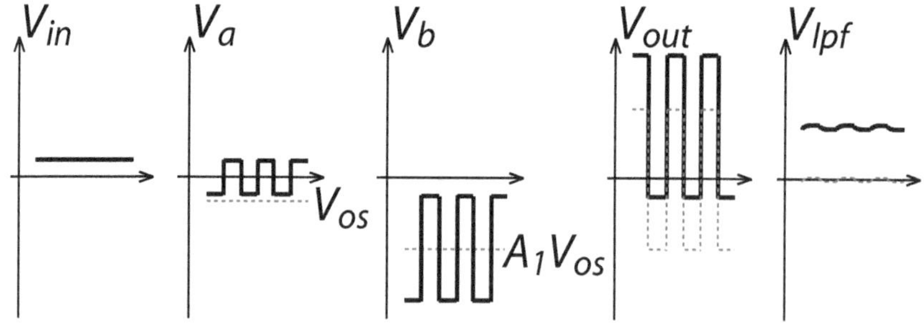

Fig. 2-10 *Idealised waveforms of a chopper amplifier in the time domain (gain=A_1=3).*

amplifier and the –3dB frequency of the LPF. The LPF can also be implemented with a sample-and-hold [2.7].

2.3.1 Noise in chopper amplifiers

When a low-pass filter is used after the chopper amplifier, the chopper ripple and low-frequency noise can be filtered. CMOS amplifiers usually have a high DC input offset voltage and *1/f* noise. To reduce the *1/f* noise, the modulation or chopper frequency chosen should be higher than the *1/f* corner frequency [2.1]. This is illustrated in figure 2-11. In this figure it is shown that the signal is modulated, and that noise and offset is superposed on this modulated signal. Afterwards the signal, noise and offset are modulated. This means that the signal is demodulated, after which the signal can be filtered out.

From this analysis it can be concluded that the chopper technique completely reduces the *1/f* noise when the chopper frequency is higher than the corner frequency of the *1/f* noise. In practice the noise level of a chopper amplifier is slightly higher than the thermal noise level. However, the need to suppress the chopper ripple means that only a low bandwidth is obtainable with a chopper amplifier.

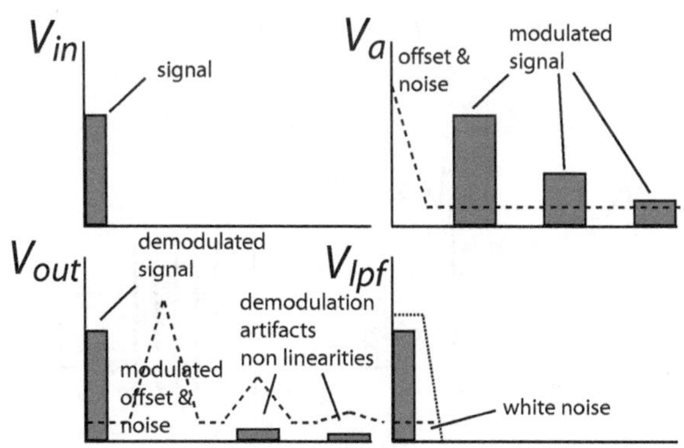

Fig. 2-11 *Idealised waveforms in the frequency domain of a chopper amplifier.*

2.3.2 Chopped operational amplifier in a feedback network

It is helpful in the understanding of chopper amplifiers to examine a chopped operational amplifier in a feedback network more closely. In figure 2-12, a chopped operational amplifier A_1 is depicted [2.8]. Resistors R_1 and R_2 are used as feedback resistors, CH_1 and CH_2 are the modulators and LPF the low-pass filter, as discussed in the previous section. The idealised waveforms are depicted in figure 2-13.

In this analysis it is assumed that the gain and bandwidth of the operational amplifier are infinite. The input of A_1 is at virtual ground because of the infinite open loop gain. Thus, the offset voltage V_1 is visible at the output V_d of the input chopper CH_1. This offset is modulated towards the input V_c of chopper CH_1, therefore it is actually visible as a square wave where, in the case of a normal operational amplifier, a virtual ground would

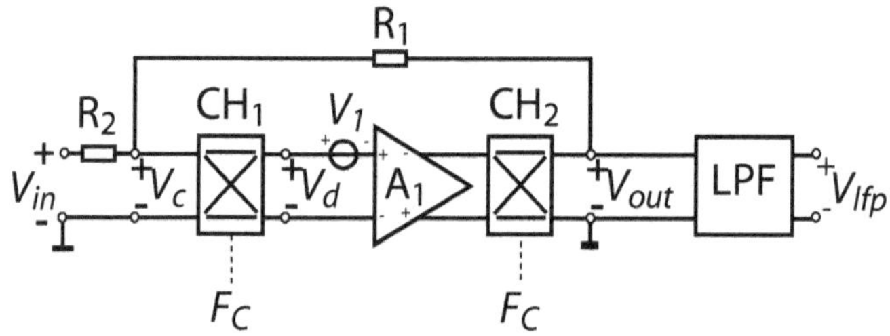

Fig. 2-12 *Chopped operational amplifier in a feedback network.*

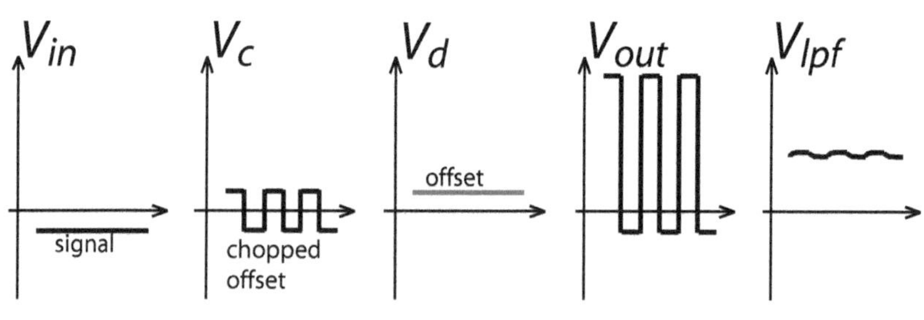

Fig. 2-13 *Idealised waveforms in the time domain of a chopped operational amplifier in a feedback network for (gain = 3).*

be expected. The output V_{out} of the chopper amplifier can now be expressed by:

$$V_{out} = V_c - \frac{V_{in} - V_c}{R_2} R_1 = -V_{in}\frac{R_1}{R_2} + V_c\left(1 + \frac{R_1}{R_2}\right). \qquad (2\text{-}17)$$

This means that the operational amplifier ripple at the output has an effective amplitude of offset V_1 times the feedback gain factor.

The conclusion which can be drawn from this analysis is that the chopper ripple of a chopped operational amplifier in a feedback network is visible at the input of the first chopper. This insight will be used in the explanation of chopper offset-stabilized chopper amplifiers discussed in section 3.4.

2.3.3 Charge injection effects in chopper amplifiers

The charge injection of chopper amplifiers on a differential chopper amplifier will be discussed. In these amplifiers the residual offset is mainly caused by the charge injection mismatch from the clock lines to the input and output of the amplifier that is chopped. In figure 2-14 this is modelled with cross talk capacitors C_1 to C_4. The resistors R_1 and R_2 model the on-resistance of the switches used in the chopper and the source resistance.

First the effect of the mismatch between C_1 and C_2 will be analysed. When both lines are loaded with identical cross talk capacitors, no residual offset will occur since it will be a common-mode spike. However, if there is a slight mismatch between the two capacitors, a differential component will also appear at V_a, which will translate into a residual offset, because these spikes are actually demodulated by the input chopper towards the input [2.6].

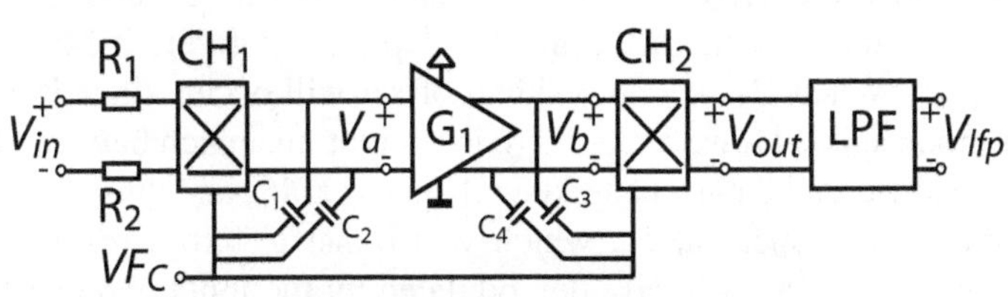

Fig. 2-14 *Charge injection model in a chopper amplifier.*

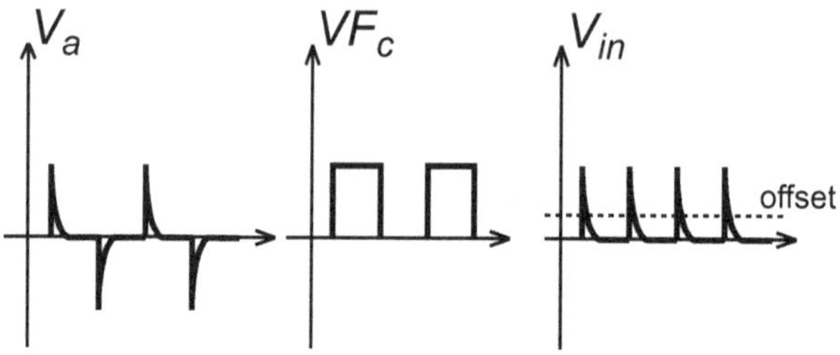

Fig. 2-15 *Illustration of residual offset caused by demodulated spikes caused by a mismatch between C_1 and C_2.*

This effect is illustrated in figure 2-15.Therefore, each time the chopper clock switches charge is being injected into the input, this differential charge can be expressed by:

$$q_{inj} = (C_1 - C_2)V_F \ , \tag{2-18}$$

where V_F is the driving voltage of clock VF_C. This charge is applied two times the per clock period. This means that a current will run through the resistors R_1 and R_2. So the residual offset can be expressed by:

$$V_{os,res1} = 2(R_1 + R_2)(C_1 - C_2)V_F F_C \ , \tag{2-19}$$

where F_C is the chopper frequency. This means that the residual offset increases with increasing chopper frequency. However, the chopper frequency needs to be higher than the $1/f$ corner frequency to obtain optimal noise performance.

For example the residual offset per unit of capacitance for a 20 kHz chopper frequency, an on-resistance of 5 kΩ with no source impedance and a 5 V driving voltage, would be 2 μV/fF.

Secondly the effect of the mismatch between C_3 and C_4 will be analysed. Also the capacitors C_3 and C_4 depicted in figure 2-14 can cause residual offset. When C_3=C_4 no residual offset will occur, since it will be a common-mode spike. However, if there is a slight mismatch between the two capacitors, a differential current spike will appear at V_b, and thus a differential voltage spike will appear at V_a, which will translate into a residual offset, because these spikes are actually demodulated by the input chopper towards the input [2.6]. This effect is illustrated in figure 2-16.

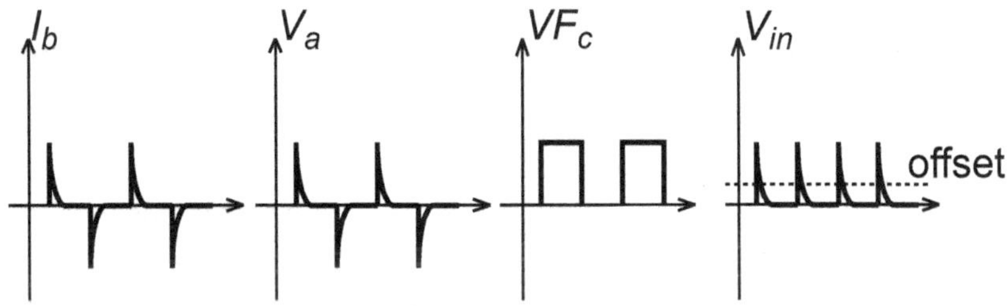

Fig. 2-16 *Illustration of residual offset caused by demodulated spikes caused by a mismatch between C_3 and C_4.*

This leads to an extra residual offset which can be expressed by:

$$V_{os,res2} = \frac{2(C_3 - C_4) V_F F_C}{G_1}, \tag{2-20}$$

where G_1 is the transconductance of the chopped amplifier. It can be seen that higher transconductance amplifiers will be less vulnerable to the mismatch of C_3 and C_4. For example for a 20 kHz chopper frequency, a 100 μA/V transconductance, and a 5 V driving voltage the residual offset per unit of capacitance would be 2 μV/fF. The residual offset caused by the charge injection can be expressed as:

$$V_{os,res} = V_{os,res1} + V_{os,res2}. \tag{2-21}$$

This implies that effort has to be put into the layout of differential choppers and clock lines when designing a chopper amplifier, because the metal to metal capacitance of signal lines can be in the order of fF's.

2.4 Chopped auto-zeroed amplifier

As already discussed in section 2.2.1, a fully integrated amplifier with three cascaded auto-zeroed amplifiers with output offset storage has been reported [2.9]. Subsequently, these stages were also chopped [2.3]. This implementation achieved an input referred offset of less than 5 μV. Theoretically, the combination of auto-zeroing and chopping would have a better offset performance because the residual offset of the auto-zero amplifier is being chopped. Practically in some applications the charge injection of the chopper is the dominant source of residual offset, and auto-zeroing will only have a small effect on the residual offset.

However, there is also an effect of the noise, because the folded white noise can be modulated as well. This reduces the effect of folded white noise. In figure 2-8 it is shown that an auto-zero amplifier with a 50% duty cycle has a minimum in the noise PSD at two times the auto-zero frequency. This means that if an auto-zero amplifier running with a 50% duty cycle is chopper modulated with the double auto-zero frequency, that the noise around DC would then be optimal. This has been sketched in figure 2-17. Note that in an implementation, the chopper and the switches driven by F_1 can be combined. The reduction in noise at low frequencies as well as the rise in noise towards the chopper frequency is sketched in figure 2-18.

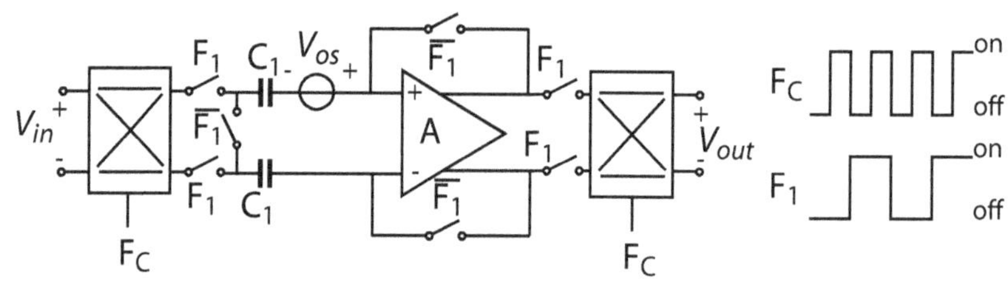

Fig. 2-17 *A chopped auto-zero amplifier, with an auto-zero duty cycle of 50% and a chopper frequency two times higher than the auto-zero frequency.*

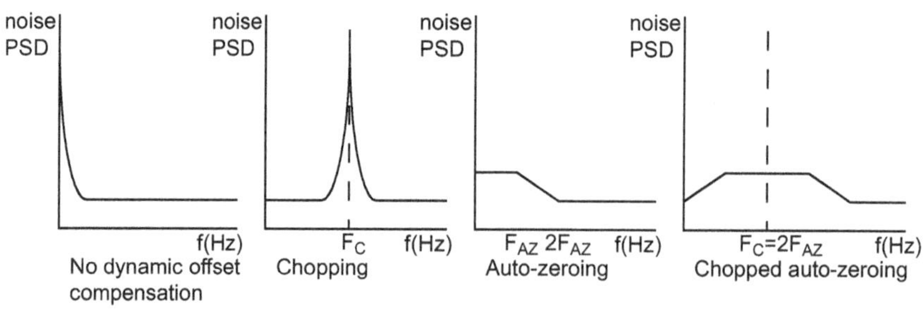

Fig. 2-18 *General output PSD of different kind of dynamic offset compensation techniques [2.10].*

The folded noise can be modulated to even higher frequencies by using a higher chopper frequency. During each signal phase the polarity of the chopper should be positive and negative for an equal time, to average out the noise contribution, that is sampled during each auto-zero phase. Therefore, optimal noise at low frequencies will be achieved when:

$$F_C = \frac{n}{d}F_1 \text{ and } n = 1,2,3\,... \qquad (2\text{-}22)$$

The technique of using both an auto-zero and a chopper technique was used in [2.10].

2.5 Switching non-idealities

All the dynamic offset compensation techniques presented in this work have one thing in common: CMOS transistors are used as switches. Therefore, the non-ideal behaviour of CMOS switches needs to be discussed.

An ideal switch is an element that does or does not allow a signal through depending on the driving signal. In other words, when it is open or off the impedance is infinite, and when it is closed or on the impedance is zero. Furthermore, there is no delay between the driving signal and the switch action.

In reality a CMOS switch has a non-infinite impedance R_{off} when it is off and a non-zero R_{on} impedance when it is on. This R_{off} is typically 100 MΩ while R_{on} can be as high as 10 kΩ for minimum size switches. A voltage drop will thus occur when current is flowing through an open switch. In these cases the R_{on} needs to be taken into account.

Secondly, there is a small time delay between the signal driven to the gates of the switch and the switching action. The main cause of delay is the relatively big capacitance of the clock line. To avoid delay time differences between lines it is better to balance the clock-line capacitances, i.e. to make clock lines the same size.

However, the main problem for dynamic offset compensation circuits is the so-called charge injection. This charge injection is caused by two phenomena: parasitic capacitive feed-through and the redistribution of channel charge.

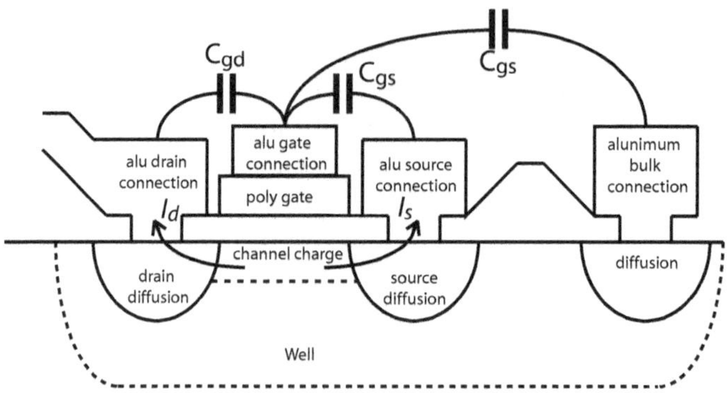

Fig. 2-19 *Charge injection model.*

In figure 2-19 the channel charge injection and parasitic capacitive feed-through are modelled. When the transistor is open, a layer of minority carriers exists under the gate between the source and drain of the transistor. This charge can be expressed for a NMOS switch as:

$$q_{inj,ch} = (V_{gs} - V_T)WLC_{ox} \ . \qquad (2\text{-}23)$$

For a minimum size switch in a 0.7μm process, the values that can be found are W=1 μm, L=0.7 μm, V_T=0.7 V and V_{gs}=5.5 V C_{ox}=2 fF/μm^2, which leads to $q_{inj,ch} = 6.72\,\text{fC}$. This charge can cause a 6.72mV voltage step on a 1pF capacitor.

This channel charge has to go somewhere when the transistor is turned off. Depending on the structure of the switch and the loads at the drain or source, it will partly flow into the load of the drain and partly into the load of the source. The charge currents I_{injd} and I_{injs} disturb the drain and the source respectively.

The parasitic capacitive feed-through can be modelled by gate-to-source C_{gs}, gate-to-drain C_{gd}, and gate-to-bulk capacitances C_{gb}, which means that the gate signal not only drives the on or off state of the transistor but it also disturbs the drain, source and bulk of the transistor. The gate-source capacitance charge injection can be expressed by:

$$q_{inj,C} = \Delta V_{gs} C_{gs} \ , \qquad (2\text{-}24)$$

where ΔV_{gs} is the voltage swing over the parasitic capacitance. Note that C_{gs} is the sum of all capacitances from driving (gate) lines to signal (source) lines. Both the capacitive feed-through and the redistribution of channel charge effects have a linear dependency on V_{gs}. This also means that changes in the source and gate voltage have an effect on the residual offset, which implies that charge injection is also a limit to the DC power supply rejection ratio (PSRR) as well as the DC common-mode rejection ratio (CMRR).

From a designer's point of view the two effects are not distinguishable from each other since they both happen at the moment of opening or closing of the transistor. Therefore, they are both considered to be charge injection. In this work some practical layout issues are discussed in appendix A to avoid disastrous clock feed-through effects.

2.5.1 Charge injection reduction tactics

As already mentioned in the analysis of the residual mismatch due to mismatch in section 2.2, an effective method to reduce residual offset caused by charge injection is to use bigger capacitors as auto-zero capacitors. Another similar way is to minimize the charge injection by using minimum size transistors as switches. However, in some applications a minimum size switch will have a too high R_{on}. In this section some other ways to cope with the charge injection challenge are presented.

Dummy switches

Charge injection of a main transistor (M_1 in figure 2-20) can be removed by means of a second dummy transistor (M_2 in figure 2-20). Generally, there are two ways to implement a dummy switch [2.11]: It can be connected in series with the switch or in parallel. This is shown in figure 2-20, where C_H is modelled to be charged with the charge injection of the switches.

In figure 2-20a this is done by putting a second transistor of half the width in series with the main transistor. The idea is that, when closing M_1, half the charge injection would go towards C_H and V_{out}, and that the other half would go towards V_{in}. A transistor of half the size driven by the opposite clock signal would remove the charge injection going into C_H.

Unfortunately, because of asymmetry in the layout of the switch and the unequal load impedances at the drain and source, the assumption of equal splitting of charge between source and drain is generally not true [2.11].

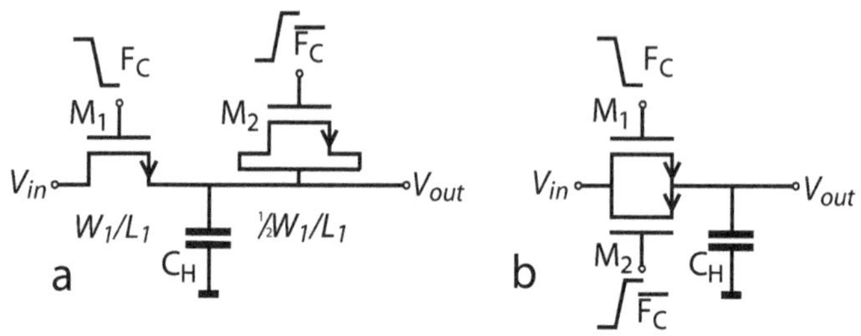

Fig. 2-20 *Charge injection cancellation of a NMOS switch with dummy switches. (a) Half width NMOS in series (b) PMOS in parallel.*

In figure 2-20b a second transistor is used in parallel to the main transistor. The idea is that while the main transistor M_1 closes, also the second transistor M_2 closes, and that the PMOS charge injection consisting of holes compensates for the charge injection of the NMOS switch consisting of electrons. The PMOS transistor in figure 2-20b will, in first order, only remove the channel charge injection component at one V_{in} voltage. This voltage can be expressed by solving the following equation:

$$(V_{FC}-V_{in,opt}-V_{T,n})WLC_{ox} = -(-V_{in,opt}-V_{T,p})WLC_{ox} \quad , \tag{2-25}$$

$$\text{which leads to } V_{in,opt} = \frac{V_{FC}-V_{T,n}-V_{T,p}}{2} \quad . \tag{2-26}$$

Furthermore, the parasitic capacitive feed-through overlap capacitances of NMOS and PMOS are not equal. As a result, the total charge injection effect will differ from equation (2-25). In conclusion, the use of dummy switches will never totally cancel charge injection. Nevertheless, it does reduce charge injection.

Differential circuits

Another way to deal with charge injection is to use differential circuits. The charge injection in a fully differential structure is, first of all, a common-mode issue. Only the charge injection mismatch will cause a differential signal. This has already been discussed section 2.3.3.

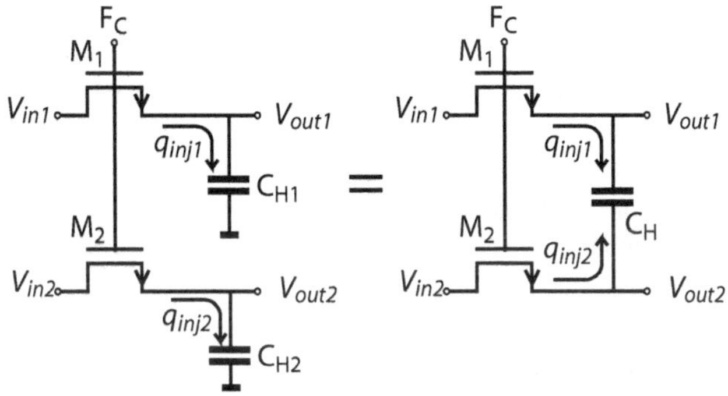

Fig. 2-21 *Differential sampling circuit with charge injection.*

In figure 2-21 a differential sampling circuit is depicted. In this figure when both q_{inj1} and q_{inj2} are equal, they do not have a consequence on the differential charge and thus do not influence the differential voltage over C_H. Only the differential charge injection or charge injection mismatch will have an effect. This charge injection mismatch is expected to be much smaller than the absolute charge injection.

From equation (2-23) it can already be concluded that a differential input voltage would already lead to a charge injection mismatch, since the V_{gs} of M_1 is not equal to the V_{gs} of M_2 [2.11]. This charge injection can be expressed as:

$$\Delta q_{inj,Vin} = q_{inj1} - q_{inj2} = \mathrm{WLC_{ox}}(V_{in2} - V_{in1}) \ . \qquad (2\text{-}27)$$

This also implies that an auto-zero or chopper amplifier with a higher offset has an increased charge injection problem. Therefore, a combination of dynamic offset compensation techniques would lead to lower residual offsets.

Mismatch of the parasitic capacitances from gate lines to source lines is critical. Any systematic parasitic capacitance due to layout asymmetry would lead to residual offset. This has already been analysed for a chopper amplifier in section 2.3.3.

When equation (2-23) is investigated, it can be concluded that the W, L, C_{ox} and threshold mismatch leads to channel charge mismatch. Mismatch characteristics of minimum size or small MOS switches are usually not available. However, if the charge injection of a MOS transistor is proportional to the area [2.12], then it can be assumed that the mismatch is proportional to the square root of the area. This would imply that minimum size switches are the best choice to minimize channel charge injection.

The use of dummy transistors in differential dynamic offset compensated circuits does not reduce residual offset, because dummy switches will only compensate for the absolute charge injection, which causes common-mode charge injection. However, dummy switches do cause more charge injection mismatch, because their use increases the effective area of a switch, which causes more residual offset.

Fixed voltage swing

The most straight forward approach is to use digital signals to operate the switches of dynamic offset compensated circuits. However, in equations (2-23) and (2-24) it can be seen that the charge injection is dependent on the voltage swing with which the switch is driven.

This would mean that if the switch were driven with a digital signal, then the DC PSRR and DC CMRR would roughly be limited from around 100–120 dB. There are ways to implement circuits that limit the voltage swing driving the switches independently from the power supply voltage. In this book the implementation of a dynamic offset compensated current-sense amplifier is described in section 6.3. The input common-mode (CM) voltage V_{inCM} ranges from 2 to 28 V. The input PMOS switches are driven from V_{inCM} to V_{inCM}–2 V by a level shift circuit. This technique achieves a 143 dB CMRR over the full input CM range and it even achieves a 152 dB CMRR when the CM ranges from 5 to 28 V. The same amplifier also has a reference input, which has a common-mode range from 0 to VDD-1.5 V. This input is modulated with a normal digital signal from 0 to VDD. The CMRR from this input voltage is 120 dB.

2.5.2 Charge injection suppression circuits

In this section three circuits will be discussed, in which charge injection reduction techniques are used. These techniques are: nested chopping, spike filtering, and the use of dead band. All these techniques have been implemented in chopper amplifiers.

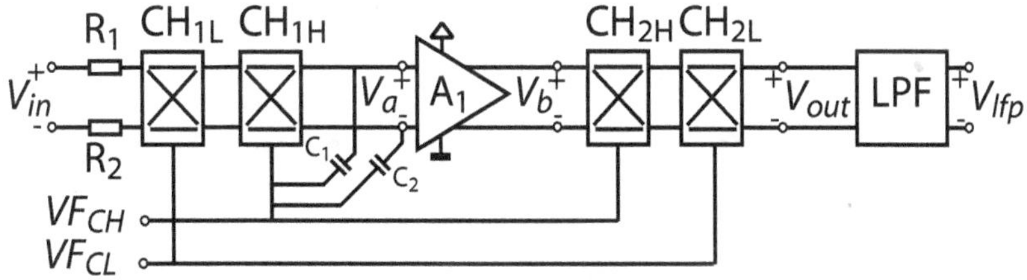

Fig. 2-22 *Nested chopper amplifier principle.*

Nested chopper

From equation (2-19) it can be seen that the residual offset due to mismatch is proportional to the chopper frequency. In the nested chopper technique this residual offset is also chopped.

In figure 2-22 the nested chopper amplifier principle is shown. The inner choppers CH_{1H} and CH_{2H} run at a frequency F_{CH}. This frequency can be chosen optimally for noise, i.e. F_{CH} should be chosen higher than the *1/f* corner frequency. This high frequency will lead to a relatively high residual offset. The nested choppers CH_{1L} and CH_{2L} modulates this residual offset at a low frequency F_{CL}, which strongly reduces the offset. The frequency F_{CL} can be chosen optimal for the input signal. Thus, the nested choppers reduces the charge injection of the inner choppers. With this technique a 100 nV offset was obtained with F_{CH}=2 kHz and F_{CL}=15,6 Hz [2.13].

This amplifier now has two ripples at the output. One ripple is caused by the offset of the amplifier A_1, which is modulated with the frequency F_{CH}. The other ripple is caused by the residual offset, which is modulated at frequency F_{CL}. To filter this ripple a LPF is needed with a -3 dB frequency smaller than F_{CL}. This makes this technique useful in low frequency applications, like custom sensor read out electronics, such as implementations of fully integrated Hall-sensors [2.14] and temperature sensors [2.15] [2.16], but not for general purpose amplifiers.

Spike filtering

When it is assumed that charge injection causes residual offset, and that charge injection is an effect which occurs during the switching action of a switch, then it can also be assumed that the charge injection causes voltage spikes with high frequency. It can be assumed that the spikes have spectral

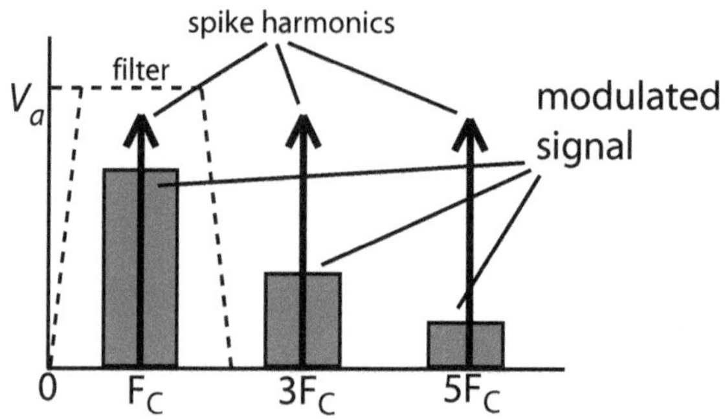

Fig. 2-23 *Modulated signal and spike harmonics.*

components on the odd harmonics of the chopper frequency [2.5]. This is sketched in figure 2-23, which implies that the spikes can be filtered by making the amplifier A_1 in figure 2-9 selective for specific frequencies.

An amplifier using this technique is illustrated in figure 2-24. A low-pass filter was used in [2.17]. A thorough analysis of this technique can be found in [2.5], in which it is concluded that the optimal choice is a second order low-pass or band-pass filter with a cut off frequency of two times the chopper frequency. An implementation with a band-pass filter can be found in [2.1]. This implementation was improved in [2.18], where a band-pass filter was matched with the chopper frequency generator so that the input offset voltage could be reduced to 0.54 μV.

This technique leads to amplifiers having a bandwidth equal to the chopper frequency or less. This is useful for custom amplifiers in sensor read out circuits.

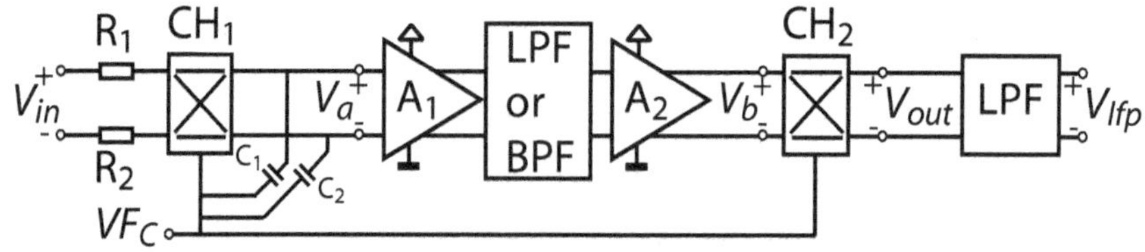

Fig. 2-24 *Chopper amplifier with a frequency selective amplifier to improve residual offset by filtering spikes.*

Delayed modulation and guard band

Instead of putting effort into filtering the spikes in the frequency domain, it is also possible to filter the spikes in the time domain. In figure 2-25a, a modulated amplifier is shown consisting of a modulator M, amplifier A_1, and demodulator D.

In figure 2-25b, a delayed demodulation scheme is shown. The idea is that when the shape of the voltage spike is known, a fixed delay can be used to allow the demodulated spike to have an average value of zero. Thus, the spike is used to compensate for itself. This has been successfully implemented [2.19] by using a low-pass filter behind the amplifier to shape the spikes. It achieved a 1 μV offset with a 6 kHz modulator frequency. However, a mismatch between the shape of the spike and the delay still causes residual offset.

Another approach, which is simpler to implement, is shown in figure 2-25c. In this dead-band or guard-band approach, the output voltage is not modulated while the spike is present. This technique has been used in [2.20] [2.21] and [2.14] for custom read out electronics for sensors. In one implementation [2.21], an amplifier with an average 200 nV input offset was realised.

In [2.14] a nested chopper technique was used. This implementation used a guard-band for the inner chopper and a partially digitally-implemented

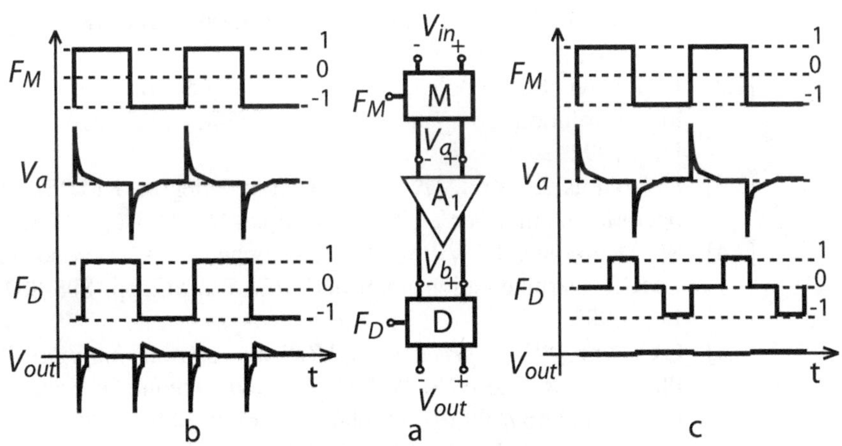

Fig. 2-25 *(a) Modulated amplifier, (b) delayed demodulation clock diagram, (c) dead band clock diagram.*

nested chopper to obtain a 3.65 μT 3σ offset Hall sensor, which is equivalent to an input voltage offset well below 200 nV.

2.6 Conclusions

In this chapter, two offset compensation techniques have been described: Auto-zeroing and chopping. The first one can be described as a time domain technique, the latter as a frequency domain filtering technique. They lead to a significantly different noise performance for low-frequency noise.

In addition, switch non-idealities of MOS transistors were discussed, since both techniques use MOS switches. The major cause of residual offset is charge injection, and design tactics as well as a couple of circuit-level to minimise the charge injection were discussed. Making use of differential circuits seems the best design tactic to reduce charge injection effects.

2.7 References

[2.1] C.C. Enz, G.C. Temes "Circuit techniques for reducing the effects of op-amp imperfections: autozeroing, correlated double sampling, and chopper stabilization", *Proc. IEEE*, pp. 1584–1614, Nov. 1996.

[2.2] B. Razavi, B.A.Wooley, "Design techniques for high-speed, high-resolution comparators", *IEEE JSSC*, pp. 1916–1926, Dec. 1992.

[2.3] R. Poujois J. Borel, "A low drift fully integrated MOSFET operational amplifier", *IEEE JSSC*, pp. 499–503, Aug. 1978.

[2.4] M. Degrauwe, E. Vittoz, I. Verbauwhede, "A micropower CMOS-instrumentation amplifier", *IEEE JSSC*, pp. 805-807, June 1985.

[2.5] C.C. Enz, *"High precision CMOS micropower amplifiers"*, Ph.D. Thesis, These No. 802, Ecole Polytechnique Federale de Lausanne, http://library.epfl.ch/en/theses/?nr=802, 1989.

[2.6] A. Bakker, J.H. Huijsing, *"High-accuracy CMOS smart temperature sensors"*, Dordrecht: Kluwer, 2000.

[2.7] A. Bakker, J.H. Huijsing, "A CMOS chopper opamp with integrated low-pass filter", *Proc. ESSCIRC*, pp. 200–203, Sep. 1997.

[2.8] K.C. Hsieh, P.R. Gray, D. Senderowicz, D.G. Messerschmitt, "A low-noise chopper-stabilized differential switched-capacitor filtering technique", *IEEE JSSC*, pp. 708–715, Dec. 1981.

[2.9] R. Poujois B. Baylac, D. Barbier, J. Ittel, "Low-level MOS transistor amplifier using storage techniques", *IEEE ISSCC*, pp. 152–153, Feb. 1973.

[2.10] A.T.K. Tang, "A 3μV-offset operational amplifier with 20nV/$\sqrt{\text{Hz}}$ input noise PSD at DC employing both chopping and autozeroing", *IEEE ISSCC*, pp. 386–387, Feb. 2002.

[2.11] B. Razavi, "*Design of analog CMOS integrated circuits*", McGraw-Hill, International edition, 2001.

[2.12] M.J.M. Pelgrom, A.C.J. Duinmaijer, A.P.G. Welbers, "Matching properties of MOS transistors", *IEEE JSSC*, pp. 1433–1439, Oct. 1989.

[2.13] A. Bakker, K. Thiele, J.H. Huijsing, "A CMOS nested-chopper instrumentation amplifier with 100-nV offset", *IEEE JSSC*, pp. 1877–1883, Dec. 2000.

[2.14] J.C. van der Meer, F.R. Riedijk, E. van Kampen, K.A.A. Makinwa, J.H. Huijsing, "A fully integrated CMOS hall sensor with a 3.65μT 3σ offset for compass applications", *IEEE ISSCC*, pp. 246–247, Feb. 2005.

[2.15] M.A.P. Pertijs, K.A.A. Makinwa, J.H. Huijsing, "A CMOS smart temperature sensor with a 3σ inaccuracy of 0.1°C from –55°C to 125°C", *IEEE JSSC*, pp. 2805–2815, Dec. 2005.

[2.16] K.A.A. Makinwa, M.F. Snoeij, "A CMOS temperature-to-frequency converter with an inaccuracy of less than 0.5°C 3σ from 40°C to 105°C", *IEEE JSSC*, pp. 2992–2997, Dec. 2006.

[2.17] C.C. Enz, E.A. Vittoz, F. Krummenacher, "A CMOS chopper amplifier", *IEEE JSSC*, pp. 335–342, June 1987.

[2.18] C. Menolfi, Q. Huang, "A fully integrated, untrimmed CMOS instrumentation amplifier with submicrovolt offset", *IEEE JSSC*, pp. 415–420, Mar. 1999.

[2.19] C. Menolfi, Q. Huang, "A chopper modulated instrumentation amplifier with first order low-pass filter and delayed modulation scheme", *Proc. ESSCIRC*, pp. 54–57, Sep. 1999.
[2.20] A. Bilotti, G. Monreal, "Chopper-stabilized amplifiers with a track-and-hold signal demodulator", *IEEE Trans. Circuits and Systems I*, pp. 490–495, Apr. 1999.
[2.21] Q. Huang, C. Menolfi, "A 200nV offset 6.5nV/$\sqrt{Hz}$ noise PSD 5.6kHz chopper instrumentation amplifier in 1μm digital CMOS", *IEEE ISSCC*, pp. 362–363, Feb. 2001.

Dynamic Offset Compensated Operational Amplifiers

3

3.1 Introduction

Typical sensor output signals are in the microvolt range and have bandwidths ranging from DC up to a few kilohertz. Boosting such signals to levels compatible with typical analog-to-digital converters requires low-offset operational amplifiers with gain-bandwidth (GBW) products of a few megahertz. For example, implementing an amplification of 40 dB with a gain accuracy of 1% and a bandwidth of 1 kHz calls for a low-offset operational amplifier with a GBW of at least 10 MHz. Achieving such GBW products in combination with a microvolt-level input offset voltage is not straightforward.

In the previous chapter, two dynamic offset compensation techniques were discussed: chopping and auto-zeroing. Chopping is a frequency modulation technique, which requires low-pass filters in the signal path to filter chopper ripple out. Therefore, chopping alone is not suitable for broadband applications. Auto-zeroing is a time domain technique in which the offset is measured and afterwards subtracted from the signal. Therefore, this technique alone is not suitable for continuous-time operation. However, in this chapter it will be shown that, by using these techniques in multi-path topologies, broadband continuous-time amplifiers can be realized.

3.2 Ping-pong operational amplifier

As described in the previous chapter, the auto-zero technique is not directly suitable for use in a continuous-time general purpose amplifier. During the sampling of the offset, it is necessary to take the amplifier out of the signal path. Therefore, implementations of amplifiers have been developed which sample the offset during start-up or at the request of a user [3.1] [3.2]. This has the advantage that there is no signal aliasing effect, because during operation there is no switching action. However, there is also no effect on $1/f$ noise and offset drift due to temperature changes during operation.

Another way to make a continuous-time amplifier by using the auto-zero technique is the ping-pong technique [3.3]. A ping-pong auto-zero operational amplifier uses two auto-zero amplifiers, which run in parallel to each other. While one is being auto-zeroed the other one is used to amplify the signal (figure 3-1). This way a continuous-time broadband operational amplifier is created by time-domain multiplexing two auto-zero amplifiers. One of the amplifiers is always present to allow accurate and stable feedback. This technique was used to implement a rail-to-rail input [3.4]. This technique was also implemented with a SAR ADC and digital offset storage [3.5].

A combination of this ping-pong technique together with the chopper technique has been used to obtain a 20 nV/√Hz noise PSD for <1 kHz operational amplifier [3.6] [3.7]. A disadvantage of the ping-pong technique is that spikes are caused because the voltages V_{b1} and V_{b2} at the output of the

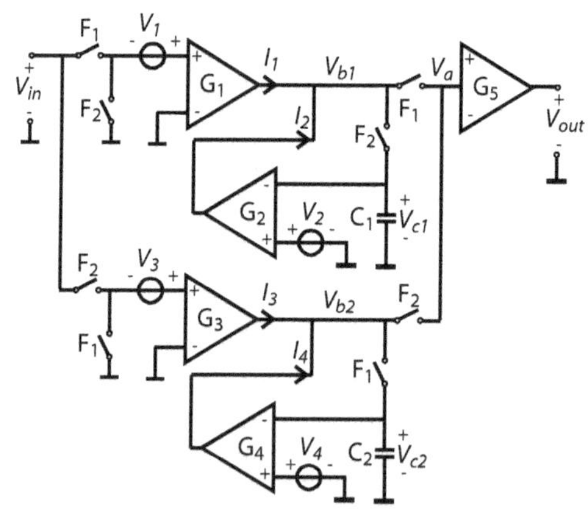

Fig. 3-1 *Concept of a ping-pong amplifier.*

input amplifiers have to switch between the offset compensating voltages V_{c1} and V_{c2} and the voltage required at the input of the output amplifier V_a. This results in spikes at the output. This effect can be reduced by replacing C_1 and C_2 with active integrators with the same input common-mode voltage as the output stage G_5. Nevertheless, spikes still remain because switching occurs within the signal path. This effect can be reduced by using a multi-path technique in which a signal path does not have any switches. These techniques will be discussed in the next section.

Implementations of ping-pong amplifiers with a multi-path structure are also commercially available [3.8]. In these amplifiers the ping-pong amplifier works in a low-frequency path and a high-frequency path is never disconnected from the signal path [3.9].

3.3 Offset-stabilized amplifiers

Offset-stabilization is another technique with which a dynamic offset compensated broadband amplifier can be designed. In this technique an auxiliary amplifier is used to compensate for the offset of a main amplifier which is never disconnected from the signal path. In figure 3-2 the basic concept of offset-stabilization is sketched. A main operational amplifier G_m with an offset V_m, is being offset-stabilized by a stabilizing amplifier G_n with a hypothetical offset of 0 V. Stage G_{m2} acts as an auxiliary input of the main amplifier. Stabilizing amplifier G_n applies a voltage to the inputs of G_{m2}, which drives a current to the output of G_m to compensate for its input offset

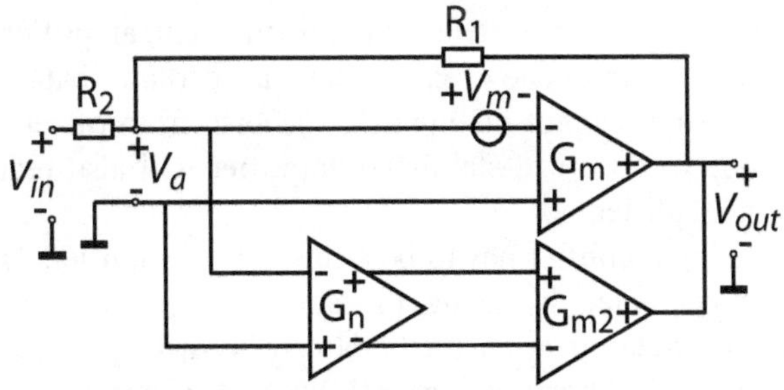

Fig. 3-2 *Offset-stabilized amplifier concept, where G_n compensates for the offset G_m via G_{m2}.*

voltage. The resistors R_1 and R_2 have been added to indicate a feedback network

It is important to note that this technique only works in negative feedback amplifiers. In a feedback configuration the differential input voltage V_a of G_n, is substantially equal to the offset V_m. The amplifier G_n measures this voltage and drives it to zero by applying a signal to G_{m2}, that acts as the auxiliary input of G_m. The residual offset due to finite gain of the combined amplifier can then be expressed as:

$$V_{os,res,gain} \approx \frac{A_m}{A_n A_{m2}} V_{os}, \tag{3-1}$$

where A_n, A_m, and A_{m2} are the DC voltage gains of the stabilizing amplifier G_n, the main amplifier G_m, and the auxiliary input of the main amplifier G_{m2}, respectively. It can be concluded from this that the combined voltage gain of the stabilizing amplifier and the auxiliary input of the main amplifier has to be much larger than the voltage gain of the main amplifier. Reducing a 10 mV worst-case offset voltage to a 1 μV residual offset, while the voltage gain of auxiliary input of the main amplifier is ten times lower than the voltage gain of the main amplifier itself, would require a minimum voltage gain of the stabilizing amplifier of 100 dB.

This topology can also be seen as a multi-path amplifier in which the cascade of the stabilizing amplifier and the auxiliary input of the main amplifier form the high-gain low-frequency path, and the main amplifier itself is the high-frequency path. Low-frequency characteristics will, therefore, be determined by the stabilizing amplifier. In that case, for instance, the low frequency noise is determined by the noise of the stabilizing amplifier G_n, while the unity gain frequency is determined by the main amplifier G_m.

Since the low-frequency path has a certain bandwidth, the offset-stabilization loop cannot distinguish $1/f$ noise from offset. Therefore, the stabilizing loop of an offset-stabilized amplifier will also reduce the $1/f$ noise of the main amplifier.

The stabilizing amplifier has to be a low offset amplifier. Therefore, it can either be implemented by using the auto-zero or chopper technique. Therefore, two offset-stabilization methods can be distinguished: auto-zero offset-stabilization and chopper offset-stabilization. Combinations of these techniques can also be used. These will be discussed in the next sections, and implementations of offset-stabilized amplifiers are presented in chapters 5 and 6.

3.3.1 Auto-zero offset-stabilized amplifiers

A diagram of a system using auto-zero offset-stabilization is shown in figure 3-3. During phase F_2 the stabilizing amplifier G_n samples its own offset voltage. In this phase stabilizing amplifier G_n measures its own offset voltage and stores an offset compensating voltage on capacitor C_1. According to the theory discussed in section 2.2.3, the residual offset due to finite gain of G_n can then be expressed as:

$$V_{osn,res,gain} = \frac{V_n}{1+A_{n2}}, \tag{3-2}$$

where A_{n2} is the DC gain of G_{n2}, which acts as an auxiliary input of the stabilizing amplifier G_n. During phase F_1, the stabilizing amplifier measures the offset voltage of the main amplifier G_m, and stores an offset compensating voltage on capacitor C_2. The resulting residual input referred offset due to finite gain can be expressed as:

$$V_{os,res,gain} \approx \frac{A_m}{A_n A_{m2}} V_m \frac{V_n}{1+A_{n2}}, \tag{3-3}$$

where A_n, A_m, A_{n2} and A_{m2} are the DC voltage gains of the stabilizing amplifier, the main amplifier, and the gains from their auxiliary inputs stages

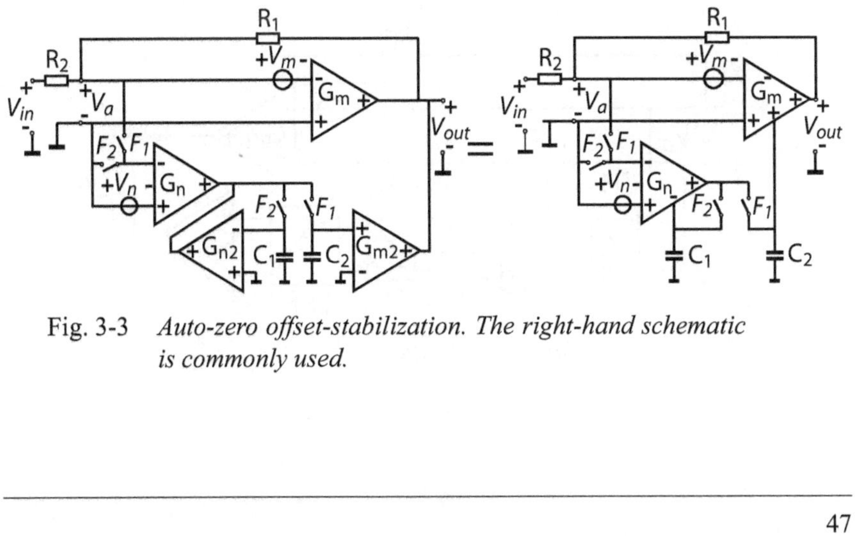

Fig. 3-3 *Auto-zero offset-stabilization. The right-hand schematic is commonly used.*

G_{n2} and G_{m2}. Compared to the ideal circuit, an additional source of residual offset is the finite gain A_{n2} and the offset V_n of the stabilizing amplifier.

Furthermore, the noise aliasing associated with auto-zeroing will give rise to a higher input-referred noise voltage at low frequencies, as discussed in section 2.2.4 [3.10]. A detailed empirical analysis of the noise behaviour of these kind of amplifiers can be found in [3.11]. Another drawback of this circuit is that, as with all capacitive auto-zeroed techniques, the charge injected by the switches is also stored on the offset compensating capacitors C_1 and C_2. Therefore, these capacitors need to be relatively large. Another drawback is that the open-loop gain of the whole amplifier is significantly different in the phases F_1 and F_2. This can be a cause of aliasing for signals above the clock frequency. Probably for this reason multi-path ping-pong amplifiers have also been implemented [3.8] [3.9].

In the literature this topology is known as chopper stabilized [3.12], [3.13] or continuous-time auto-zero [3.10]. Devices with this topology are commercially available [3.12] [3.14], and achieve typical input offset voltages of 1 μV. This topology is also used in a device which could operate up to 200°C [3.15].

3.3.2 Chopper offset-stabilized amplifiers

Some so-called chopper-stabilized amplifiers [3.12] [3.13] use auto-zeroed stabilizing amplifiers and are, therefore, auto-zero offset-stabilized amplifiers. A true chopper offset-stabilized amplifier, which uses a chopper compensation amplifier, has been proposed in [3.16]. A chopper offset-stabilized amplifier has been presented [3.17] using only 17 μA of supply current.

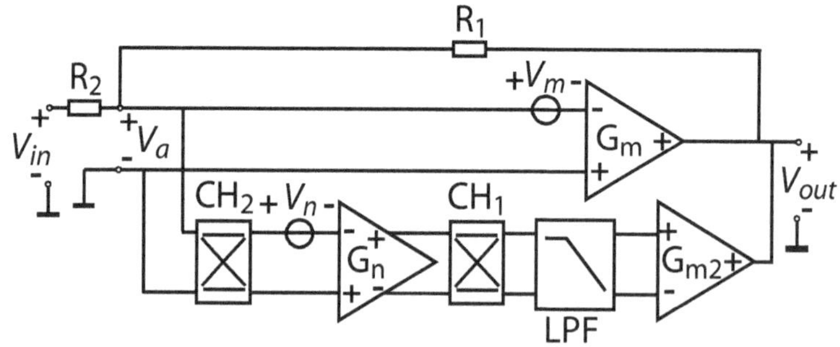

Fig. 3-4 *Chopper offset-stabilization.*

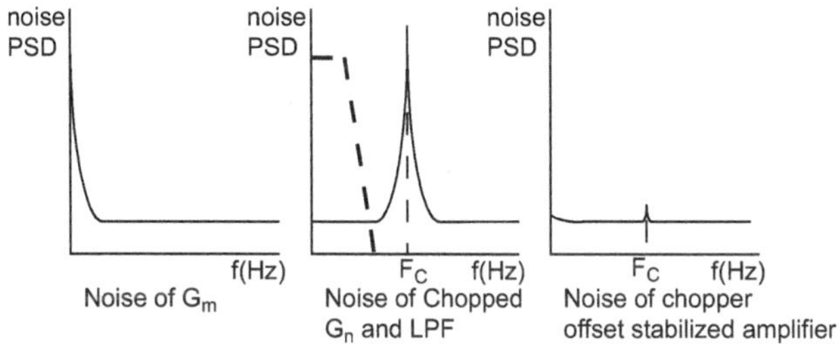

Fig. 3-5 *Noise in chopper offset stabilized amplifiers.*

In the chopper offset-stabilized amplifier shown in figure 3-4, the chopper amplifier composed of chopper CH_2, stabilizing amplifier G_n, and chopper CH_1, senses the offset of the main amplifier G_m. A low-pass filter LPF removes the chopper residuals that are caused by the offset V_n of stabilizing amplifier G_n. The residual offset due to finite gain is then expressed by equation (3-1).

The effects of noise in chopper offset stabilized amplifiers is sketched in figure 3-5. The offset and $1/f$ noise of the stabilizing amplifier itself are modulated to the chopping frequency F_c, and are removed by the low pass filter. For effective suppression of $1/f$ noise, the bandwidth of the stabilizing loop as well as the chopper frequency should, therefore, be larger than the main $1/f$ noise corner frequency of the main amplifier. Thus, for optimal noise behaviour, the –3 dB frequency of the LPF should be chosen higher than the $1/f$ corner frequency of the main amplifier G_m, and the chopper frequency should be high enough to properly filter out the chopper residuals.

When the bandwidth of the stabilizing loop and the chopper frequency are chosen optimally, the white noise level of the stabilizing amplifier G_n will be equal to the low frequency noise. In contrast, the white noise at low frequencies of a similarly dimensioned auto-zero offset-stabilized amplifier would be significantly higher due to noise aliasing.

In conclusion, compared to the use of auto-zero offset-stabilization, the use of chopper offset-stabilization should lead to better performance with respect to noise and offset. Therefore, this topology is used in the implementations discussed in chapter 5.

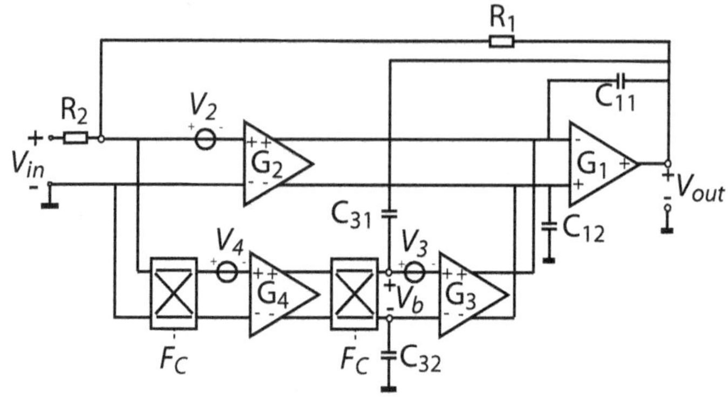

Fig. 3-6 *Conceptual chopper offset-stabilized multi-path nested Miller compensated operational amplifier.*

3.3.3 Frequency compensation

In figure 3-6 a chopper offset-stabilized amplifier is depicted as a multi-path amplifier [3.16]. In this circuit, G_2 and G_1 form a high-frequency path, while G_4, G_3, and G_1 form a chopped low-frequency path. Stage G_3 is effectively the auxiliary input of G_2. The Miller capacitors C_{11}, C_{12}, C_{31}, and C_{32} form a multi-path nested Miller frequency compensation network [3.18]. A problem with this circuit is that the offset V_4 of G_4 together with the choppers give rise to a chopper ripple. The circuit itself lacks a low-pass filter except for the Miller capacitors, which limit the unity gain frequency of the whole amplifier.

An improved design of this circuit is shown in figure 3-7. An integrator composed of transconductor G_5 and capacitors C_{51} and C_{52} has been added to the low frequency loop. This has the advantage that the integrator acts as a low-pass filter and limits the chopper ripple. The circuit shown in this figure can be considered a multi-path amplifier in which the cascaded transconductances G_2 and G_1 form the low-gain/high-frequency path, while the cascaded transconductances G_4, G_5, G_3, and G_1 form the high-gain/low-frequency path. Instead of multi-path nested Miller compensation, multi-path hybrid-nested Miller compensation has been implemented [3.19] [3.20]. If the topologies are designed to have a smooth

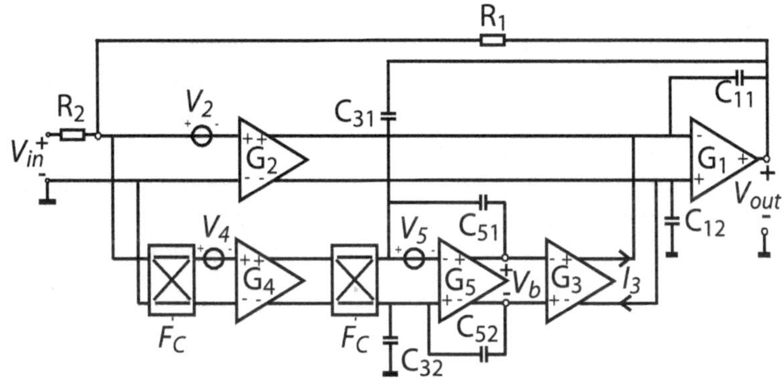

Fig. 3-7 *Chopper offset-stabilization using an active integrator and multi path hybrid-nested Miller compensation.*

roll-over of both paths, then the unity gain frequency of both topologies can be given by:

$$f_{0dB} = \frac{G_4}{2\pi C_3} = \frac{G_2}{2\pi C_1} , \tag{3-4}$$

where C_3 and C_1 are the values of capacitors C_{31} and C_{32}, and, C_{11} and C_{12}, respectively. The modulated offset V_4 of G_4 is now filtered by the integrator. However, it still gives rise to chopper ripple in the form of a triangular wave at the output V_b of the integrator, which in turn gives rise to a triangular wave at the input of the whole amplifier. The input referred peak-to-peak voltage of this triangular wave of the topology shown in figure 3-7 is given by:

$$V_{in\text{-}pp} = \frac{V_4 G_4 G_3}{2 F_C C_5 G_2} , \tag{3-5}$$

where F_C is the chopper frequency and C_5 is the value of the integrator capacitors C_{51} and C_{52}.

Compared to the compensated high-frequency path, this path contains an extra low-frequency pole due to the presence of the integrator. Therefore, without the presence of capacitors C_{31} and C_{32}, the open loop frequency of the amplifier response will have a –40 dB/decade roll-off at low frequencies.

As a result, the amplifier is only conditionally stable. This means that in order to ensure stability without external frequency compensation, the closed loop gain must be limited to below a certain value, which is not desirable in a general purpose operational amplifier. Bode plots of the amplifier depicted in figure 3-7 are illustrated in figure 3-8.

The topology shown in figure 3-7 can be seen as a four-stage low-frequency path and two-stage high-frequency path. The frequency compensation can also be extended to more stages, as shown in figures 3-9 and 3-10. The benefit of the topology shown in figure 3-9 is that the two-stage

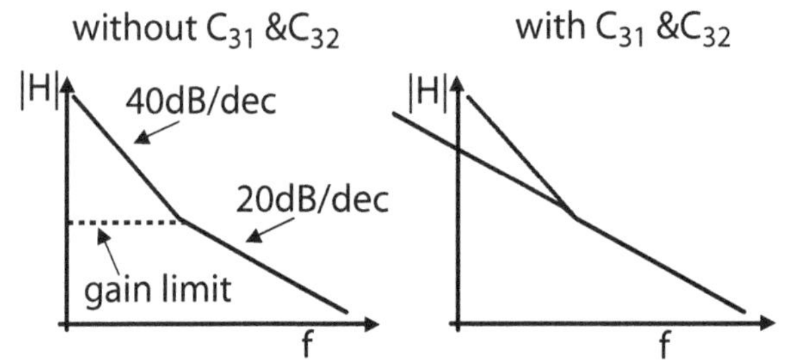

Fig. 3-8 *Sketched bode plot of the amplifier topology depicted in figure 3-7 with and without hybrid-nested Miller capacitors C_{31} and C_{32}.*

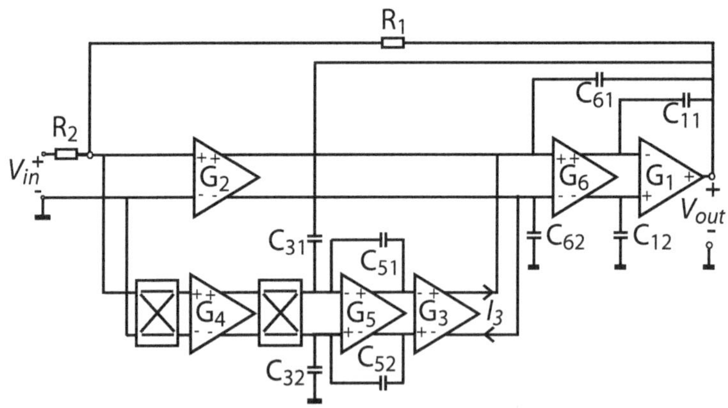

Fig. 3-9 *Chopper offset-stabilization using an active integrator and multi-path hybrid-double-nested Miller compensation.*

output stage consisting of G_6 and G_1 can be easier to implement. A benefit of a two-stage integrator consisting of G_7 and G_5 as shown in figure 3-10 is the additional voltage gain of the low-frequency path, which according to equation (3-1) leads to a lower residual offset due to finite gain.

The unity gain frequency of the topologies shown in figures 3-9 and 3-10 can be expressed by:

$$f_{0dB} = \frac{G_4}{2\pi C_3} = \frac{G_2}{2\pi C_6} = \frac{G_6}{4\pi C_1}, \qquad (3\text{-}6)$$

where C_6 is the value of capacitors C_{61} and C_{62}. The operational amplifier implementation discussed in the next chapter uses the topologies of figures 3-7 and 3-10. Referring to figure 3-7, it can be noticed that the offset voltage V_5 of the integrator appears as a square wave voltage over the capacitor C_{p4}.

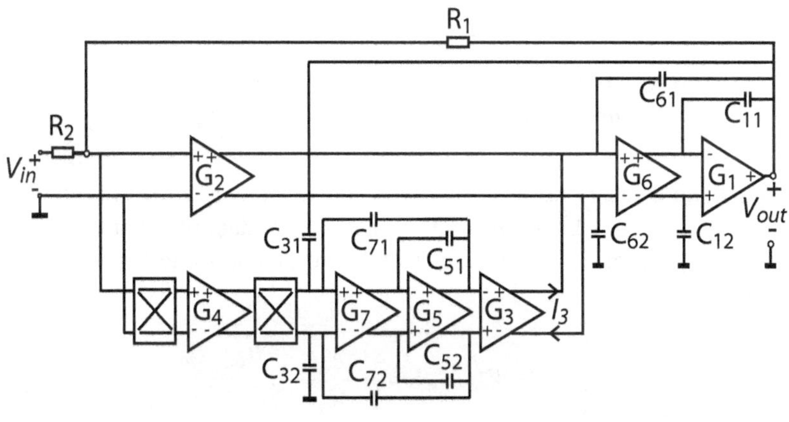

Fig. 3-10 *Chopper offset-stabilization using an active two-stage Miller compensated integrator and multi-path hybrid-double-nested Miller compensation.*

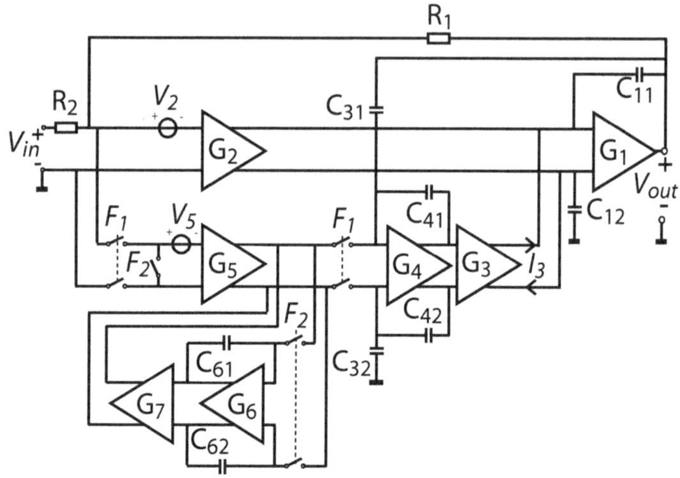

Fig. 3-11 *Auto-zero offset-stabilization using an active integrator and multi-path hybrid-nested Miller compensation.*

Charging and discharging this capacitor leads to an alternating output current at the output of G_4, which effectively leads to a residual offset at the input of the whole amplifier. The residual offset due to this parasitic capacitance can be expressed as:

$$V_{os,res,par} = \frac{4V_5 F_C C_{p4}}{G_4}. \tag{3-7}$$

This means that either the parasitic capacitance needs to be minimized or the offset of the integrator can be minimized, for instance by using nested offset-stabilization techniques, as will be shown in all the implementations presented in chapters 5 and 6.

The multi-path hybrid-nested Miller frequency compensation can also be used in auto-zero offset-stabilized amplifiers. An example of such a topology is given in figure 3-11. In phase F_2 the offset V_5 of G_5 causes a current, which is integrated by the integrator composed of G_6 and capacitors C_{61} and C_{62}. The integrator produces a voltage at the input of G_7, which acts as an auxiliary input stage for G_4 and compensates for the offset, as discussed in section 2.2.3. In phase F_1 the offset voltage V_2 is measured by G_5 and compensated for via the integrator, which is composed of G_4 and capacitors C_{41} and C_{42}, which applies a voltage to G_3. Stage G_3 acts as an offset compensating auxiliary stage. This topology works exactly like the one depicted in figure 3-3, except that the multi-path hybrid-nested Miller compensation is used by applying capacitors C_{32} and C_{31}. This version is not preferred, as the auto-zeroed amplifier stage G_5 determines the low-frequency noise, including the folded noise associated with auto-zeroing.

3.3.4 Chopper stabilized amplifiers with ripple filters

A chopper ripple is still present in the output signal of a hybrid-nested Miller compensated chopper offset-stabilized operational amplifier. This chopper ripple can be reduced by applying both auto-zeroing and chopping techniques, as will be shown in the next section.

There are also methods that use a sample and hold, which samples the chopper ripple at the output of the integrator. Two examples will be given in this section.

In figure 3-12 an implementation of the nested Miller compensated topology shown in figure 3-6 is depicted. An extra low-pass filter is implemented with a switched capacitor notch filter, which is composed of the

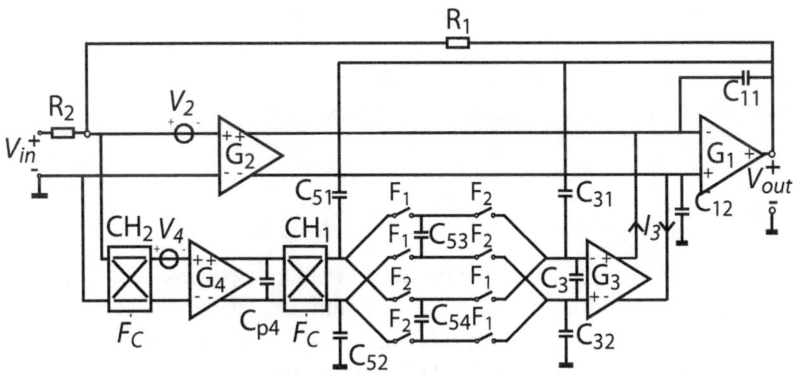

Fig. 3-12 *Chopper offset-stabilized operational amplifier with SC notch filter and synchronous integration and multi-path hybrid-nested Miller compensation [3.17].*

switches driven by F_1 and F_2 and the capacitors C_{51} and C_{52} [3.21]. This notch filter also acts as a passive integrator. To compensate for the extra pole introduced by this integrator, the capacitors C_{51} and C_{52} are introduced for the same reason as the hybrid-nested Miller compensation, as shown in the previous section. Therefore, in fact it is a hybrid-nested Miller compensated circuit. Capacitors C_{31} and C_{32} help to maintain local loop stability. The capacitor C_3 helps to limit the bandwidth of the low-frequency path such that the delay caused by the notch filter does not cause instabilities [3.17]. The advantage of this topology is that it uses a minimum of gain stages, which can lead to an amplifier that only consumes 17 μA of power [3.22] while obtaining a 55 nV/√Hz input referred white noise and a GBW of 350 kHz.

To compensate for the offset V_2 of G_2, a DC voltage has to be present at the input of G_3. This voltage is proportional to the offset V_2. A voltage at the output of chopper CH_1 together with a parasitic capacitor at the input of CH_1 would lead to a residual offset. This residual offset due to the parasitic output capacitance C_{p4} of G_4 can be expressed as:

$$V_{os,res,par} = \frac{4F_C C_{p4}}{G_4} \frac{V_2 G_2}{G_3} , \tag{3-8}$$

which means that an optimal residual offset is obtained when $G_3 > G_2$. This a strange requirement for an auxiliary input, because any switching noise of the notch filter would be amplified towards the input by a ratio G_3/G_2.

Another disadvantage of this topology is that the residual offset due to finite gain is limited by the voltage gain of the gain stage G_4. Since the residual offset due to finite gain can be expressed as:

$$V_{os,res,gain} \approx \frac{A_2}{A_3 A_4} V_2 , \tag{3-9}$$

where A_2, A_3 and A_4 are the voltage gains of the respective gain stages G_2, G_3 and G_4. The transconductances G_2 and G_3 share the same output impedance. If $G_2=G_3$, this implies that $A_2=A_3$. In this case, all offset compensation comes down to the DC voltage gain of G_4. Despite this, an implementation of this topology has a typical offset of 2 μV, and a maximum offset of 10 μV [3.22]. This is achieved with $A_2=A_3$ and A_4=86 dB, and the parasitic capacitor C_{p4} being minimized [3.17]. A disadvantage is that any switching noise of the notch filter would directly be visible at the input. This would generate a ripple since each sample would have an error caused by the kT/C noise of the sampling capacitors.

The topology shown in figure 3-13, has been implemented [3.23] and will be discussed in section 5.2. In this topology the functions of the integrator and the notch filter are not combined. The integrator is composed of G_5, C_{51} and C_{52}. The sample-and-hold circuit is composed of the switches driven by F_1 and F_2 and the capacitors C_{S1} and C_{S2}. Clearly, the benefit of this topology is that the residual offset due to finite gain can be expressed as:

$$V_{os,res,gain} \approx \frac{A_2}{A_3 A_4 A_5} V_2 \, . \tag{3-10}$$

This relaxes the voltage gain specifications of G_4. The residual offset due to the parasitic output capacitance C_{p4} of G_4 can be expressed by equation (3-7), which means that in this design the ratio A_2/A_3 can be chosen larger than 1 without compromising any residual offset. This also suppresses the kT/C noise of the sample-and-hold toward the input, achieving better ripple behaviour.

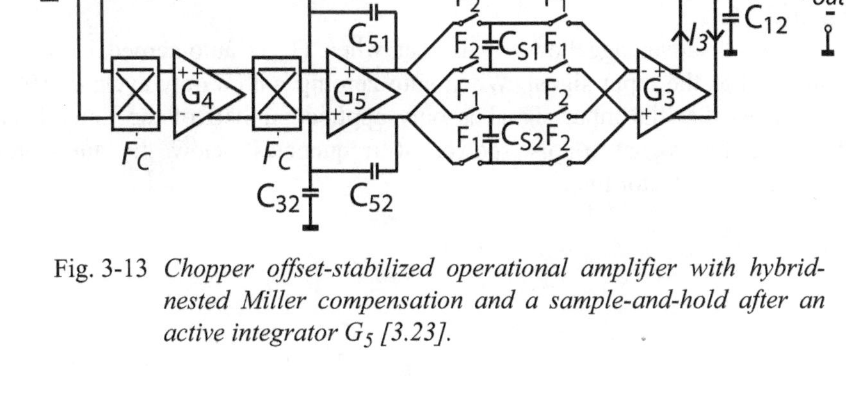

Fig. 3-13 *Chopper offset-stabilized operational amplifier with hybrid-nested Miller compensation and a sample-and-hold after an active integrator G_5 [3.23].*

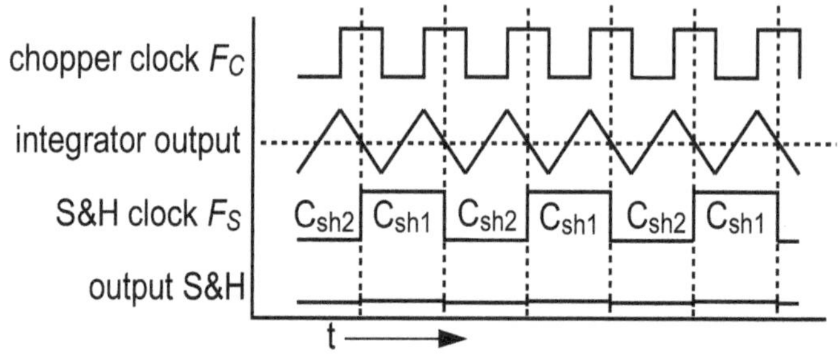

Fig. 3-14 *Clock diagram of the topology depicted in figure 3-13.*

A possible clock diagram is depicted in figure 3-14. In this diagram the chopper ripple is sampled at half the chopper frequency and with a quarter chopper period delay. With this clock scheme the output range of the integrator is used optimally.

3.3.5 Chopper and auto-zero stabilized amplifiers

Another possibility which leads to a strongly reduced chopper ripple is to auto-zero the chopper amplifier. This has been shown in the topology depicted in figure 3-15. During phase F_1 the chopper amplifier G_4 is auto-zeroed by the integrator composed of G_6 and C_{61} and C_{62}. While in phase F_2 the chopper amplifier is in the loop.

This implementation is quite powerful because the auto-zeroing of G_4 strongly reduces chopper ripple. While the folded noise associated with auto-zeroing is modulated to higher frequencies, as sketched in figure 2-18 in section 2.4, at higher frequencies the high-frequency path dominates the noise, and effectively filters out most of the noise. This has been sketched in figure 3-16.

An disadvantage however is that when G_4 is auto-zeroed it is not connected to the input signal. When auto-zeroing with a duty cycle of 50% this means that the input signal is only applied for half of the time. This decreases the signal-to-noise power at frequencies below the auto-zero frequency by a factor two.

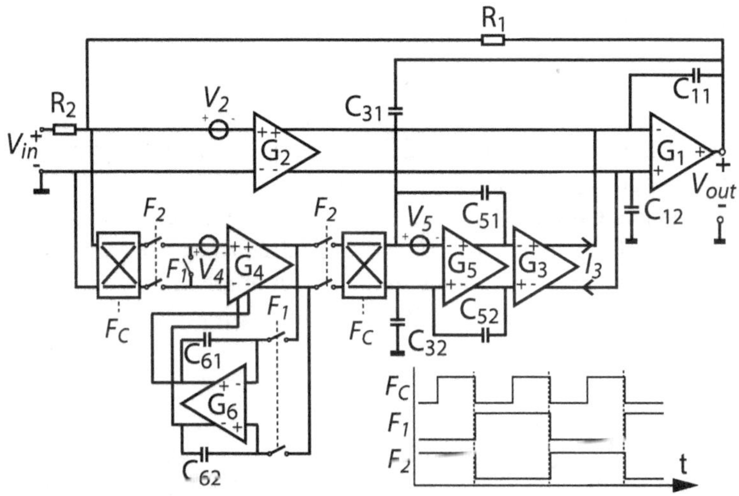

Fig. 3-15 *Chopper and auto-zero offset-stabilized operational amplifier.*

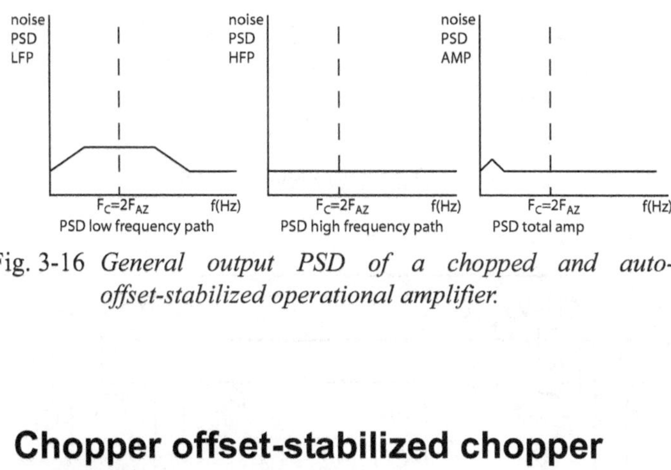

Fig. 3-16 *General output PSD of a chopped and auto-zeroed offset-stabilized operational amplifier.*

3.4 Chopper offset-stabilized chopper amplifiers

Until now the offset-stabilizing technique has been discussed. But what if an offset-stabilized amplifier itself would be chopped? Or in other words, what if a chopper amplifier would be offset-stabilized? As concluded in section 2.5.1, a dynamic offset compensated differential amplifier with a low offset has a lower charge injection problem, than that of an amplifier with higher offset. As a result, a chopper amplifier itself would benefit from a stabilized offset.

In figure 3-17 a chopper amplifier consisting of chopper CH_2, amplifier stage G_2, chopper CH_1, and output stage G_1 is shown. The offset is chopper offset-stabilized by chopper CH_4, stabilize stage G_5, chopper CH_3, an integrator around G_4, and an auxiliary input stage G_3.

The choppers CH_2 and CH_4 can be combined by placing the inputs of G_5 at the input of the amplifier. In section 2.3.2 it was shown that the modulated offset appears at the inputs of a chopper amplifier. Therefore, it is not necessary to create a modulated offset two times with CH_2 and CH_4. The

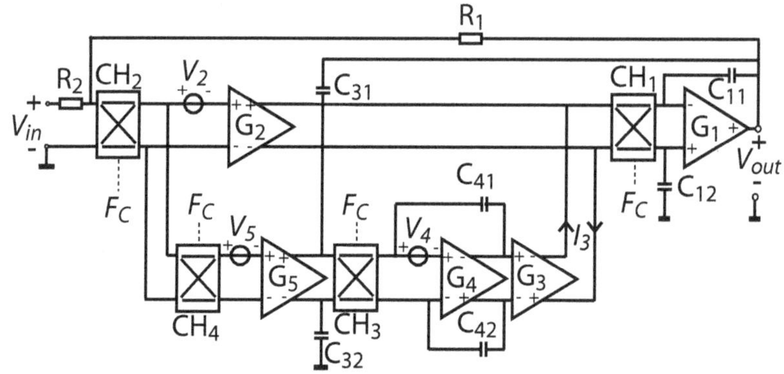

Fig. 3-17 *Chopper offset-stabilized chopper amplifier.*

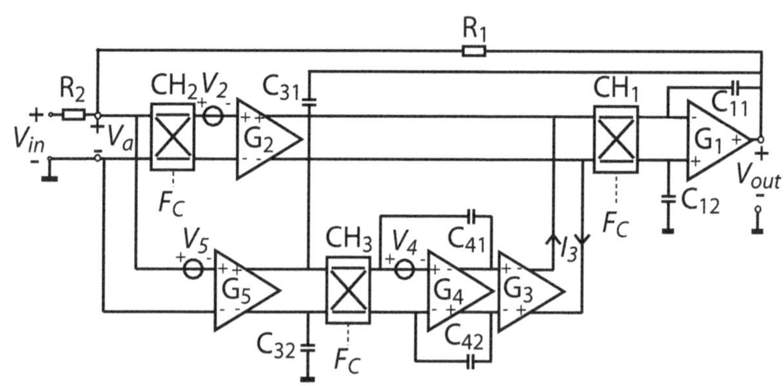

Fig. 3-18 *Chopper offset-stabilized chopper amplifier with a combined chopper CH_2.*

resulting amplifier is shown in figure 3-18 [3.24]. This benefits the residual offset due to charge injection, because the amount of switches is reduced.

The frequency behaviour is more difficult to explain than the behaviour of the offset-stabilized amplifier. Amplifier stage G_5 senses the voltage V_a at its input. This contains the modulated offset V_2 of G_2. The voltage V_a is converted into a current by G_5. This current is modulated by CH_3. The modulated current now contains DC current information of the offset V_2. Integrating this current into the integrator composed of G_4, C_{41}, and C_{42} leads to an offset reducing current I_3. However, for the input signal the combination of choppers CH_3 and CH_1 and the integrator G_4 acts as a bandpass filter around the chopper frequency. This means that the low-frequency behaviour is determined by G_2 instead of G_5.

The hybrid-nested Miller frequency compensation capacitors are still necessary to ensure that the roll-off remains 20 dB per decade around the clock frequency. The capacitors need to be placed at the left of CH_3 to provide negative feedback for the compensation capacitors.

This topology modulates the residual offset of the chopper stabilized operational amplifier depicted in figure 3-7, which would lead to a better offset specification. However, there are choppers introduced into the main signal path, and the complexity is increased.

However, the chopper ripple caused by the modulated offset of G_5 remains. Only its shape changes from a triangular wave into a saw-tooth wave, i.e. a chopped triangular wave. This ripple can again be suppressed by auto-zeroing the input stage or placing a sample-an-hold behind the integrator. A quest to obtain a low offset and low ripple resulted in the amplifier topology discussed in the next sub-section.

3.4.1 Iterative offset-stabilization

All the techniques discussed so far can be applied to operational amplifiers. Numerous combinations can be made with chopper, auto-zero and offset-stabilized amplifiers. In figure 3-19 a three-stage chopper amplifier is shown that consists of output stage G_1, intermediate stage G_2, and chopped input stage G_3.

The input stage is both chopper and auto-zero offset-stabilized. The transconductance stage G_5, and the integrator around G_4 offset-stabilized G_3, while the integrator around G_6 is used to auto-zero G_5. The offset of integrator G_4 limits the ripple reduction by crosstalk through C_{p5} to the

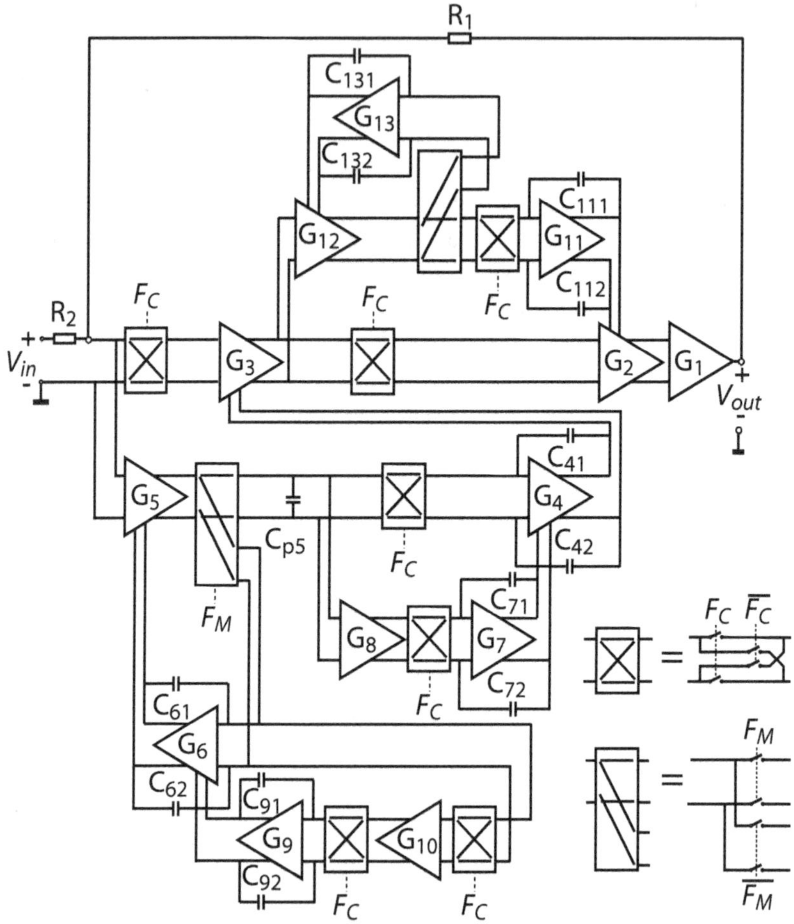

Fig. 3-19 *Chopper amplifier with iterative offset-stabilization [3.24].*

output. To circumvent this around G_4 a nested chopper offset-stabilization loop is added consisting of transconductance G_8, and an integrator around G_7. The integrator around G_6 is also chopper stabilized by transconductance G_{10}, and an integrator around G_9. The intermediate stage G_2 is also chopper and auto-zero offset-stabilized by transconductance stage G_{12}, the integrator around G_{11}, while the integrator around G_{13} is used to auto-zero G_{12} [3.24].

The resulting amplifier has 13 stages, or 19 it the necessary 6 auxiliary stages are counted. A clear disadvantage is the complexity of the topology. This topology has potentially a very low offset, since every source of offset has been offset-stabilized.

3.5 Conclusions

In this chapter several topologies have been discussed that can be used to obtain broadband dynamic offset compensated operational amplifiers. In table 3-1 the properties of the different topologies are summarized. The first topology is the ping-pong topology, where one amplifier is auto-zeroed while another amplifier is used to amplify a signal. It doesn't display a chopper ripple, although it does display switching spikes due to the switches in the signal path. Its residual offset is limited by the gain of the amplifier. The auto-zeroed offset-stabilized topology (AZOS), is less complex than the ping-pong topology, although the low-frequency (LF) noise increases by a factor $\sqrt{2}$. The chopper offset-stabilized (CHOS) topology has a superior LF noise, although it exhibits a chopper ripple. Two techniques were discussed to reduce this ripple at the cost of complexity: chopper offset-stabilization with ripple filters with a sample-and-hold (CHOS+S&H), and chopper and auto-zero offset-stabilized (CH&AZOS) operational amplifiers. A way to obtain a lower residual offset

Table 3-1. *Comparison of the different broadband dynamic offset compensated topologies.*

Topology	Ripple	LF noise	Residual offset	Complex-ity
Ping-pong	+	-	+/-	-
AZOS	++	--	+/-	+
CHOS	--	++	+	+
CHOS+S&H	+	++	+	+/-
CH&AZOS	+	+	+	+/-
CHOSCH	--	++	++	-
Iter	+	++	+++	--

is to use a chopper offset-stabilized chopper topology (CHOSCH). In this topology the residual offset of a chopper offset-stabilized amplifier is chopped, reducing the offset further. However, the topology complexity increases and the chopper switches in the signal path exhibit switching spikes. Finally the iterative (ITER) topology could lead to a very low residual offset at the cost of an increased topology complexity.

Of these topologies, the chopper offset-stabilized (CHOS) operational amplifier offers the best option for obtaining a low-noise amplifier with a reasonable topology complexity. However, chopper ripple remains a problem. Therefore, chopper offset-stabilization with ripple filters, and chopper and auto-zero offset-stabilized topologies will be used in the implementations discussed in chapters 5 and 6.

3.6 References

[3.1] F. Krummenacher, R. Vafadar, A. Ganesan, V. Valence, "A high-performance autozeroed CMOS opamp with 50μV offset", *IEEE ISSCC*, pp. 350–351, Feb. 1997.

[3.2] Texas Instruments, "Family of self-calibrating (self-cal) precision CMOS rail-to-rail output op amps (rev. B)", Datasheet TLC4501, www.ti.com, Apr. 2001.

[3.3] C.G. Yu, R.L. Geiger, "An automatic offset compensation scheme with ping-pong control for CMOS operational amplifiers", *IEEE JSSC*, pp. 601–610, May 1994.

[3.4] I.E. Opris, G.T.A.A. Kovacs, "Rail-to-rail ping-pong op-amp", *IEEE JSSC*, pp. 1320–1324, Sep. 1996.

[3.5] M. Kayal, R.T.L Saez, M. Declercq, "An automatic offset compensation technique applicable to existing operational amplifier core cell", *CICC*, pp. 419–422, May 1998.

[3.6] A.T.K. Tang, "A 3μV-offset operational amplifier with $20nV/\sqrt{Hz}$ input noise PSD at DC employing both chopping and autozeroing", *IEEE ISSCC*, pp. 386–387, Feb. 2002.

[3.7] Analog devices, "Zero-drift, single supply, rail-to-rail, input/output operational amplifier, AD8628/8629/8630", Datasheet Ref F, http://www.analog.com, Feb. 2008.

[3.8] Texas Instruments, "0.05uV/C max, single-supply CMOS op amps zero-drift", Datasheet, Rev. D, OPA335, www.ti.com, June 12, 2003.

[3.9] Thomas Kugelstadt, "Auto-zero amplifiers ease the design of high-precision circuits", Texas Instruments, focus.ti.com/lit/an/slyt204/slyt204.pdf, 2005.

[3.10] C. C. Enz, G. C. Temes, "Circuit techniques for reducing the effects of op-amp imperfections: autozeroing, correlated double sampling, and chopper stabilization", *Proc. IEEE*, pp. 1584–1614, Nov. 1996.

[3.11] I.G. Finvers, J.W. Haslett, F.N. Trofimenkoff, "Noise analysis of a continuous-time auto-zeroed amplifier", *IEEE Trans. on Circuits and Systems II*, pp. 791–800, Dec. 1996.

[3.12] Intersil, "2MHz, super chopper-stabilized operational amplifier", FN2920.10, Datasheet ICL7650S, http://www.intersil.com, Apr. 2007.

[3.13] M.C.W. Coln, "Chopper stabilization of MOS operational amplifiers using feed-forward techniques", *IEEE JSSC*, pp. 745–748, Dec. 1981.

[3.14] Analog devices, "Zero-drift, single-supply, rail-to-rail input/output operational amplifier, AD8551/8552/8554", datasheet, Ref C, http://www.analog.com, Mar. 2007.

[3.15] I.G. Finvers, J.W. Haslett, F.N. Trofimenkoff, "A high temperature precision amplifier", *IEEE JSSC*, pp. 120–128, Feb. 1995.

[3.16] C.I. Menolfi, "*Low Noise CMOS Chopper Instrumentation Amplifiers for Thermoelectric Microsensors*", 1st ed. Konstanz, Germany: Hartung-Gorre Verlag, chapter 5, 2000.

[3.17] R. Burt, J.A. Zhang, "Micropower chopper-stabilized operational amplifier using a SC notch filter with synchronous integration inside the continuous-time signal path", *IEEE JSSC*, pp. 2729–2736, Dec. 2006.

[3.18] R.G.H. Eschauzier, L.P.T. Kerklaan, J.H. Huijsing, "A 100MHz 100-dB operational amplifier with multipath nested Miller compensation structure", *IEEE JSSC*, pp. 1709–1717, Dec. 1992.

[3.19] R.G.H. Eschauzier, R. Hogervorst, J.H. Huijsing, "A programmable 1.5V CMOS class-AB operational amplifier with hybrid nested Miller compensation for 120dB gain and 6MHz UGF", *IEEE JSSC*, pp. 1497–1504, Dec. 1994.
[3.20] J.H. Huijsing, M.J. Fonderie, B. Shahi, "Frequency stabilization of chopper-stabilized amplifiers", US patent Nr. 7,209,000, Apr. 24, 2007.
[3.21] A. Bakker, J.H. Huijsing, "A CMOS chopper opamp with integrated low-pass filter", *Proc. ESSCIRC*, pp. 200–203, Sep. 1997.
[3.22] Texas Instruments, "1.8V, micropower CMOS operational amplifiers zero-drift series", Datasheet OPA333, Rev. C, May 1, 2007.
[3.23] J.F. Witte, K.A.A. Makinwa, J.H. Huijsing, "A CMOS chopper offset-stabilized opamp", *IEEE JSSC*, pp. 1529–1535, July 2007.
[3.24] J.H. Huijsing, M.J. Fonderie, "Chopper chopper-stabilized operational amplifiers and methods", US patent Nr. 6,734,723, Nov. 7, 2004.

Dynamic Offset Compensated Instrumentation Amplifiers

4

4.1 Introduction

In many sensor systems there is a need to amplify a weak differential signal, which is often accompanied by a strong common-mode signal. Amplifiers designed to handle such tasks are known as instrumentation amplifiers. There are three general approaches to design instrumentation amplifiers.

The first approach involves the use of operational amplifiers and resistive feedback. The simplest example is the differential amplifier shown in figure 4-1a. This amplifier suffers from unequal and low input impedances. Thus its gain is dependent on the source impedances. To solve this problem, the famous three operational amplifiers instrumentation amplifier has been developed, as shown in figure 4-1b [4.1]. In this topology, two operational amplifiers are used to implement a fully differential buffer that precedes a differential amplifier. This topology has a much higher input impedance but still suffers from a not very high common-mode rejection ratio (CMRR) due to resistor mismatch. Therefore, commercially available instrumentation amplifiers are often designed with the help of laser trimmed resistors. To obtain low-offset in a resistive feedback instrumentation amplifier, the operational amplifiers which are used in the implementation can be designed with dynamic offset compensation techniques, as discussed before.

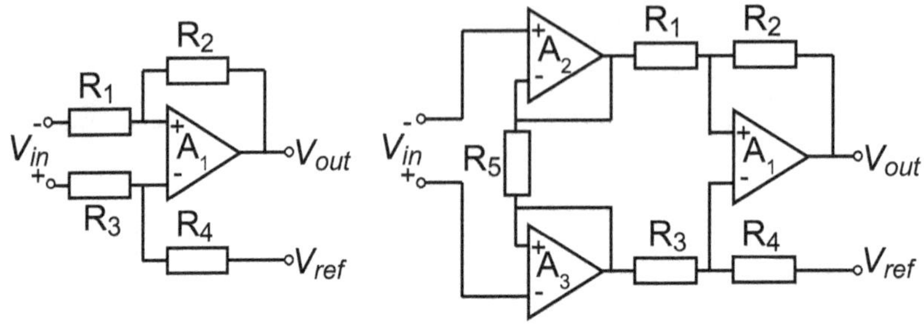

(a) *(b)*

Fig. 4-1 *(a) Differential amplifier and (b) resistive feedback instrumentation amplifier.*

The second approach involves using a flying or switched capacitor topology to overcome the large common-mode voltage [4.2] [4.3] [4.4]. A simplified version of a switched capacitor differential amplifier is shown in figure 4-2. To obtain low offset for switched capacitor circuits, the auto-zero methods can be used because the switches and capacitors are already present in the implementation.

Current-feedback instrumentation amplifiers offer another option, in which the CMRR is determined by isolation and balancing techniques. This is the subject of the next section.

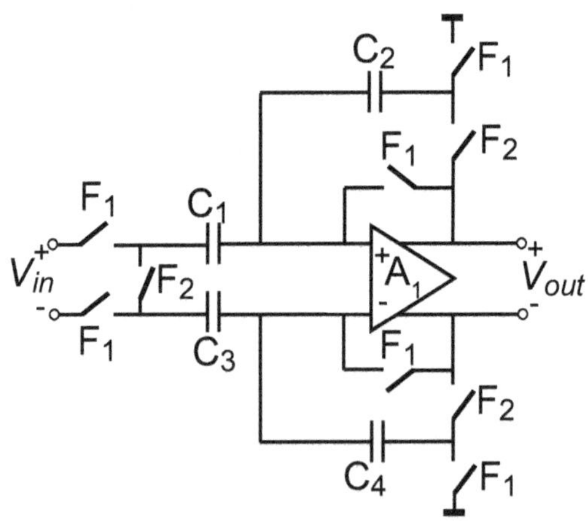

Fig. 4-2 *Switched capacitor instrumentation amplifier.*

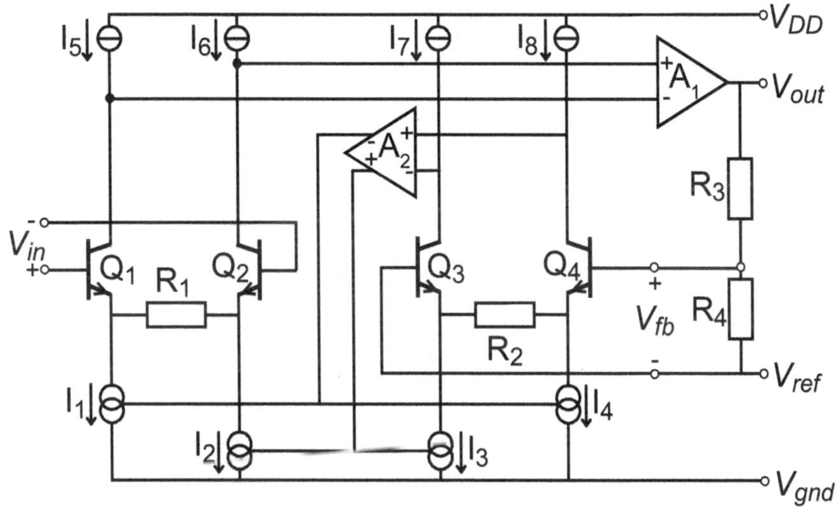

Fig. 4-3 *Current-feedback instrumentation amplifier with two feedback amplifiers.*

4.1.1 Current-feedback instrumentation amplifiers

Since the 1970s, various instrumentation amplifier topologies have been developed for current-feedback or current balancing instrumentation amplifiers. For historical reasons, these topologies are shown implemented with bipolar transistors in this section.

The first monolithic current-feedback instrumentation amplifier [4.5] used the topology shown in figure 4-3. It uses two voltage amplifiers A_1 and A_2 and two transconductance amplifiers, namely an input transconductance implemented with transistors Q_1 and Q_2 and resistor R_1, and a feedback transconductance implemented with Q_3, Q_4 and R_2.

This circuit works as follows. The input transconductance is unbalanced by the input voltage V_{in}, which causes the current through current sources I_1 and I_2 to be unequal. Voltage amplifier A_1 drives the output voltage V_{out} and consequently the feedback voltage V_{fb}. The feedback transconductance is unbalanced by this feedback voltage, which causes the current through current sources I_3 and I_4 not to be equal. Loop amplifier A_2 drives the two current mirror pairs I_1, I_4 and I_2, I_3 to be equal. The loop amplifiers A_1 and A_2 help to maintain constant currents through I_5, I_6 and I_7, I_8, respectively. By keeping the collector currents of $Q_{1,2}$ and $Q_{3,4}$ equal, their base-emitter voltages are equal. As a result the voltages over R_1 and R_2 accurately represent V_{in} and V_{fb}. Thus, the next equations hold:

$$I_1 - I_2 = 2V_{in}/R_1 , \tag{4-1}$$

$$I_4 - I_3 = 2V_{fb}/R_2 , \tag{4-2}$$

$$I_1 = I_4 \text{ and } I_2 = I_3 , \tag{4-3}$$

$$\text{Thus } \frac{V_{fb}}{V_{in}} = \frac{R_1}{R_2} \text{ and } \frac{V_{out}}{V_{in}} = \frac{R_1}{R_2}\frac{R_3 + R_4}{R_4} . \tag{4-4}$$

The settling time of this topology was rather poor, because during an overload condition, the whole amplifier needs to settle [4.6]. The feedback loop includes the input transconductance, voltage amplifier A_1, the feedback transconductance and the voltage amplifier A_2. The use of multiple feedback loops also leads to stability issues. The two topologies discussed next overcome these issues, by using only one feedback loop.

Indirect current-feedback instrumentation amplifier

The indirect current-feedback topology is shown in figure 4-4 [4.5]. It was first used [4.7] to implement a method for frequency compensation and later [4.8] to implement an instrumentation amplifier. The idea is that I_1 to I_4 are equal, and that I_5 and I_6 are equal. The loop amplifier A_1 then forces the differential collector current from Q_1 and Q_2 into Q_3 and Q_4, by driving the required voltage V_{fb} at the inputs of Q_3 and Q_4.

Another way of looking at this circuit is that the feedback transconductance amplifier, implemented with Q_3, Q_4 and R_2, together with the loop amplifier A_1, form a two-stage operational amplifier. For this circuit the frequency compensation is well-known [4.9]. A replica transconductance amplifier implemented with Q_1, Q_2 and R_1 is added to this operational amplifier and is used as a new input stage to implement the indirect current-feedback.

A disadvantage of this however, is that the collector currents of Q_1 and Q_2 and of Q_3 and Q_4 are not a linear function of V_{in} and V_{fb}, respectively.

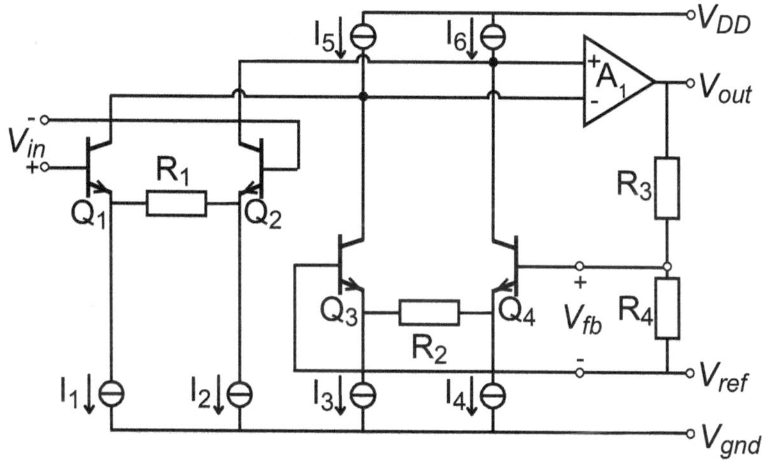

Fig. 4-4 *Indirect current-feedback.*

These non-linearities are cancelled in first order when the transistors match and $R_1 = R_2$ [4.10]. Equation (4-4) then becomes:

$$\frac{V_{out}}{V_{in}} = \frac{R_3 + R_4}{R_4} \tag{4-5}$$

The input stages can also be linearized by using gain-boosted transistors [4.8]. In CMOS design, where mismatch is a bigger problem, the use of composite transistors might be necessary to obtain a better gain accuracy and linearity [4.11].

Direct current-feedback instrumentation amplifier

For biomedical applications, which demand low power, another topology was investigated [4.12]. This topology is shown in figure 4-5. In this topology the input and feedback transconductance amplifiers are cascoded. This way the supply current is reduced. This topology was later used in CMOS implementations [4.13] [4.14]. A disadvantage of this topology in comparison with the indirect current-feedback approach is the increase in minimum supply voltage and decrease in the common-mode voltage range. Another

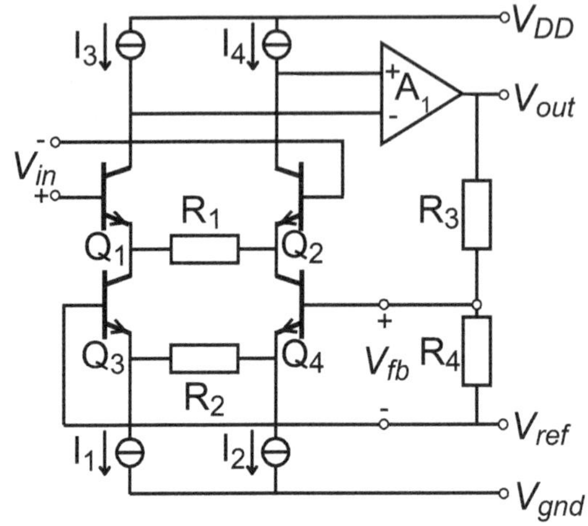

Fig. 4-5 *Direct current-feedback instrumentation amplifier.*

disadvantage is that Q_1 and Q_2 always carry the same current, whereas Q_3 and Q_4 carry a signal dependent current. This can be a cause of non-linearity.

A distinction can be made between direct and indirect current-feedback instrumentation amplifiers [4.10], where the input and feedback transconductances are cascoded and cascaded respectively. The direct current-feedback is better for low-power applications, while the indirect current-feedback approach is better for gain accuracy and for low-voltage applications, where the relatively higher common-mode input voltage range is also beneficial. The implementations presented in chapter 6 focus on wide-bandwidth offset-stabilized indirect current-feedback instrumentation amplifiers.

The indirect current-feedback instrumentation amplifier can be seen as an operational amplifier with a replica input pair. In figure 4-6 an example is given of a two-stage Miller compensated operational amplifier that consists of stages G_1 and G_2 with an additional replica input stage G_3 to implement a current-feedback instrumentation amplifier.

The gain accuracy of the instrumentation amplifier depends on the equality of G_2 and G_3. With ordinary differential pairs of the same type working in weak inversion, and well matched tail currents, a 1% gain accuracy can be obtained. The gain can be expressed as:

$$\frac{V_{out}-V_{ref}}{V_{in}}=\frac{R_1+R_2}{R_2}\frac{G_2}{G_3}. \tag{4-6}$$

A strange feature with respect to operational amplifier design is that the input stages have to be able to handle a finite input voltage, since the virtual ground concept is not applicable. However, the differential input range of ordinary differential pairs is effectively limited to about 50 mV, because the differential pairs will saturate. This input range can be improved with degeneration resistors. An implementation of an accurate transconductance amplifier is presented in section 6.3.3.

For this reason the input stages of the designs shown in figures 4-3 to 4-5 are all shown with degeneration resistors. Another odd feature is that the polarity of G_3 does not have an effect on the stability of the circuit, since G_3 is not part of the feedback network.

In figure 4-7 a multi-path current-feedback instrumentation amplifier is depicted. What is interesting is that in a multi-path current-feedback instrumentation amplifier the polarity of the input replica stages G_4 and G_6 can be chosen arbitrarily. For instance a positive gain for the low-frequency path and a negative gain for the high-frequency path. An amplifier like this can be used as an all-pass filter, since it has a phase shift of 180° between low

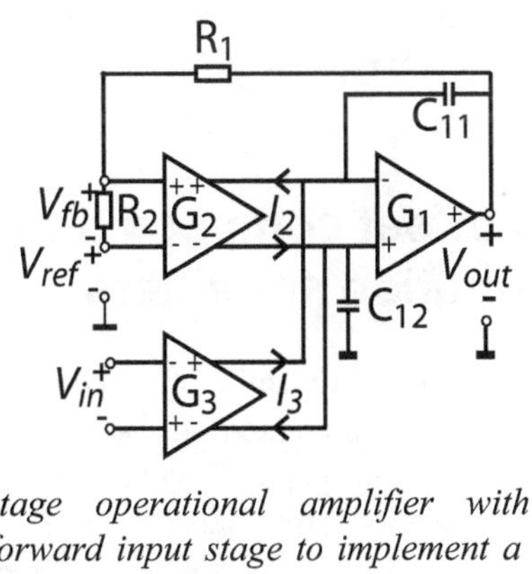

Fig. 4-6 *Two-stage operational amplifier with an additional feed-forward input stage to implement a current-feedback instrumentation amplifier.*

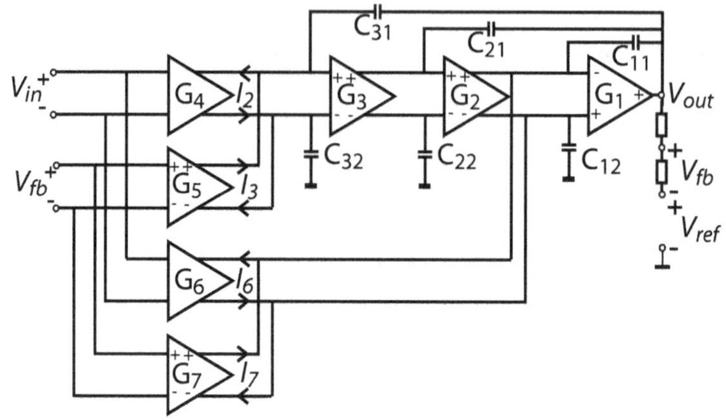

Fig. 4-7 *Four-stage multi-path current-feedback instrumentation amplifier.*

and high frequencies. However, generally this is unwanted behaviour which can surprise a designer.

The CMRR of this instrumentation amplifier is now dependent on the CMRR of the input stages which can be designed to be very high. Apart from the dependency of the offset on CM voltage, there is also common-mode voltage dependency of the voltage gain, because, the transconductances of the input stages have a common-mode voltage dependency. The offset of such an amplifier is equal to the sum of the input offset voltages of the input stage and the replica stage, this is 2–20 mV for typical CMOS implementations. The gain accuracy is limited by the matching of the input and replica transconductance which leads typically to a 1% gain accuracy.

4.2 Dynamic offset compensated instrumentation amplifiers

In this section some topologies of dynamic offset compensated current-feedback instrumentation amplifiers are presented. In section 4.2.1 the chopping and auto-zero techniques discussed in chapter 2 will be extended to the discussion of current-feedback instrumentation amplifiers. From section 4.2.3 onwards the focus will be on broadband dynamic offset compensated instrumentation amplifiers. The ping-pong amplifier discussed in section 3.2 will be extended to a discussion on ping-pong-pang amplifiers in section 4.2.3. The offset-stabilized operational amplifiers discussed in section 3.3 will be extended in section 4.2.5. The chopper offset-stabilized chopper amplifiers discussed in section 3.4 will be extended in section 4.2.6.

4.2.1 Chopper instrumentation amplifier

A chopper current-feedback instrumentation amplifier topology is shown in figure 4-8. Generally it can be assumed that the offset is two times worse than for a chopper operational amplifier, because the replica input stage also has offset and an additional chopper is also needed. The CMRR of such an amplifier is very high for signal frequencies below the chopping frequency.

As shown in section 2.5.2 the charge injection is a function of the input signal. This means that charge injection varies slightly over the input voltage range. This is a cause of gain non-linearity. For an input range of 50 mV and a decent chopper layout with minimum size switches, the non-linearity caused by this effect can be in the order of 0.1%, which is in the same order as that of the non linearity of V-to-I convertors.

The gain accuracy of the instrumentation amplifier caused by a mismatch of G_2 and G_3 can also be reduced by dynamic matching of the input stages. If there is a mismatch of Δ between G_2 and G_3, the gain error would then average over time as can be expressed by:

$$\varepsilon_{gain,mismatch} = \left(\frac{1+\Delta}{2}+\frac{1}{2+2\Delta}\right)-1 = \frac{2+2\Delta+\Delta^2}{2+2\Delta}-1 \approx \frac{\Delta^2}{2}, \qquad (4\text{-}7)$$

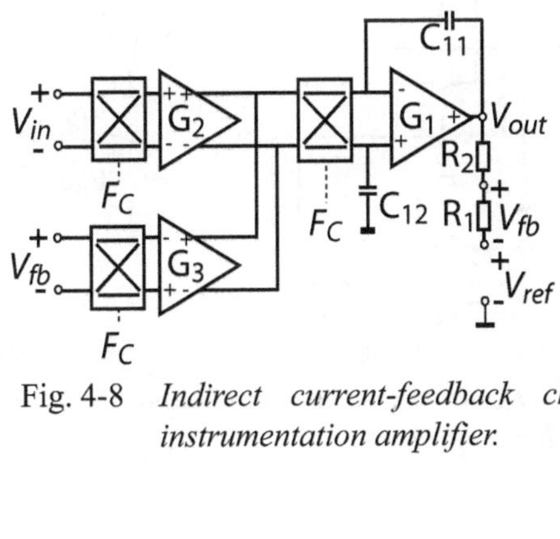

Fig. 4-8 *Indirect current-feedback chopper instrumentation amplifier.*

which would mean that a 1% mismatch would only cause a 0.005% gain error. This dynamic element matching would be relatively easy to implement, because both switches and the logic to drive them are already present. However, the inputs have to switch from the input common-mode to the reference common-mode. When those common-modes differ too much, this could cause nasty spikes, which means that the dynamic matching of the input stages cannot be applied to general purpose instrumentation amplifiers.

4.2.2 Auto-zeroed instrumentation amplifier

An auto-zeroed current-feedback instrumentation amplifier topology is shown in figure 4-9. In this figure the auto-zeroing with an auxiliary input stage G_4 is used. In phase F_1 the offset voltages V_2 and V_3 cause a current $I_{offset} = V_3G_3 - V_2G_2$. This current is integrated by the integrator around G_5 until $I_4 = I_{offset}$. Thus, after a few auto-zero cycles the offsets V_2 and V_3 will be compensated for by an offset compensating current I_4. During phase F_2 the amplifier amplifies the input signal.

An auto-zeroed current-feedback instrumentation amplifier with input offset storage is shown in figure 4-10. In phase F_1 the offsets V_2 and V_3 are stored on capacitors C_{A1} to C_{A4}. With respect to the topology shown in figure 4-9, this amplifier will probably have a higher residual offset due to the charge injection of the input switches.

However, a beneficial attribute of this amplifier is that the input common-mode voltage is separated from the inputs of the instrumentation amplifier, such as in a switched capacitor amplifier. This means that the inputs

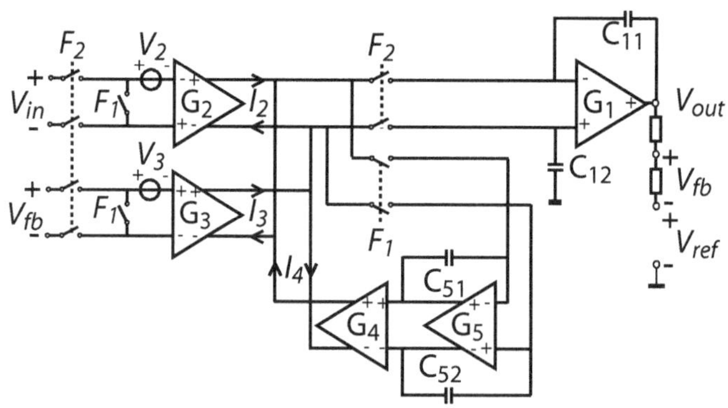

Fig. 4-9 *Auxiliary input auto-zeroed current-feedback instrumentation amplifier.*

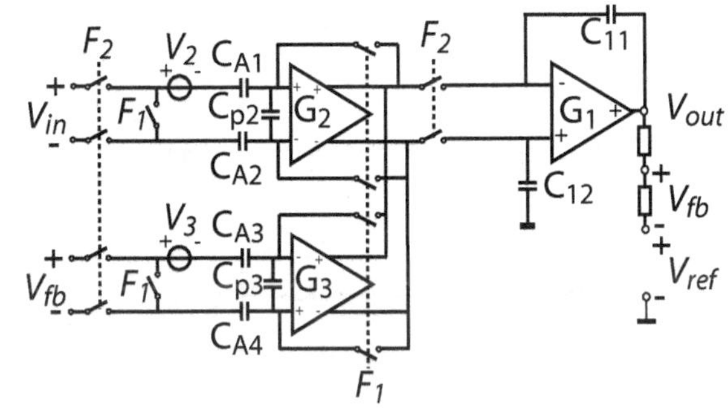

Fig. 4-10 *Input offset storage auto-zeroed current-feedback instrumentation amplifier.*

do not have to be designed to accommodate the common-mode. The dynamic element matching, as discussed before, might be easier to implement. The switches in front of the capacitors C_{A1} to C_{A2}, have to accommodate the input common-mode voltage. Consequently, this topology can be seen as a hybrid of the switched capacitor amplifier and the current-feedback instrumentation amplifier.

A disadvantage of this is, however, that the parasitic input capacitances C_{p2} and C_{p3} of transconductances G_2 and G_3 act as a voltage divider, and alter the overall gain. Effort has to be taken into matching those parasitics.

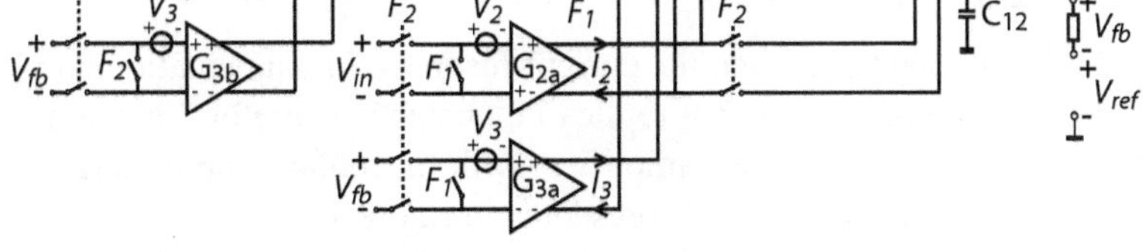

Fig. 4-11 *Ping-pong auto-zero instrumentation amplifier [4.15].*

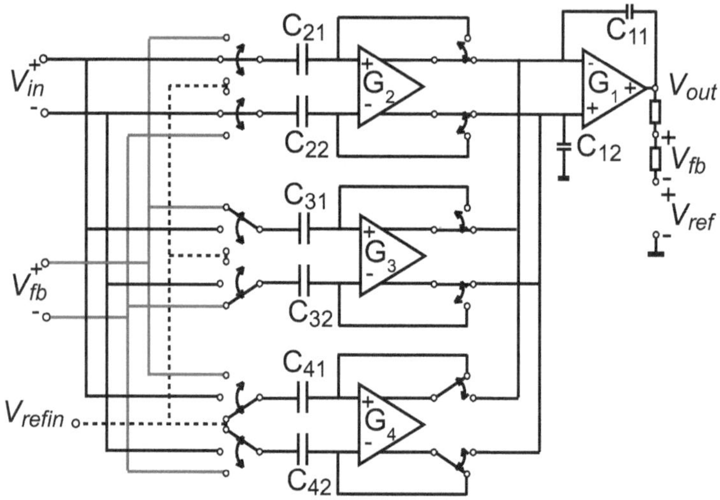

Fig. 4-12 *Ping-pong-pang auto-zero instrumentation amplifier [4.16].*

4.2.3 Ping-pong instrumentation amplifier

A wide-band instrumentation amplifier can be obtained with the ping-pong technique as described in section 3.2. The auto-zeroed amplifier as depicted in figure 4-9 can be used, by duplicating G_2, G_3, G_4 and G_5 and using the auto-zero switches as multiplexer between the two obtained input stages. An instrumentation amplifier like this is depicted in figure 4-11. A chopped ping-pong instrumentation amplifier has been used in an implementation [4.15] obtaining a 3 μV offset voltage and 140 dB CMRR.

4.2.4 Ping-pong-pang instrumentation amplifier

As was explained earlier, the current-feedback instrumentation amplifier is an operational amplifier with a replica input stage. Therefore, it can also be implemented as a ping-pong amplifier with a replica input stage. This ping-pong-pang amplifier [4.16] is presented in figure 4-12.

There are numerous algorithms that can be implemented. One example is given in figure 4-13. In this algorithm two input stages are swapped during each cycle, while each input stage is auto-zeroed with a 1/3 duty cycle at an auto-zero frequency.

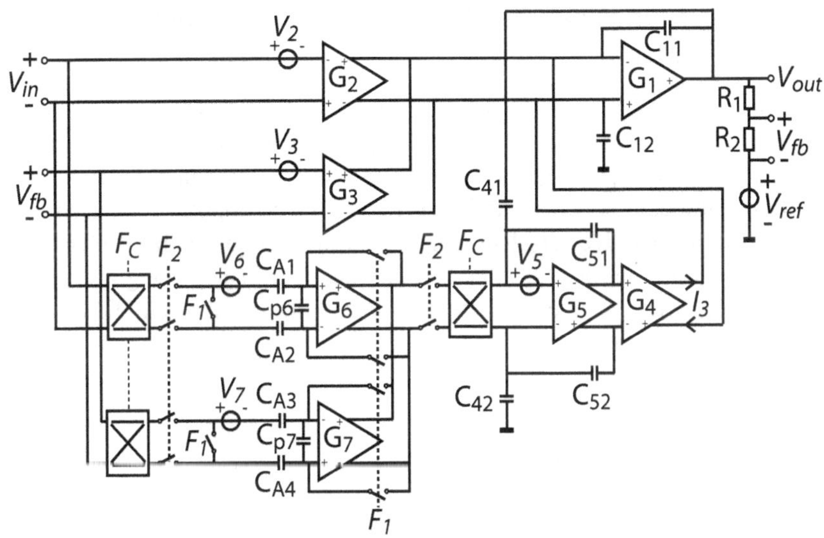

Fig. 4-16 *Chopper and auto-zero offset-stabilized current-feedback instrumentation amplifier.*

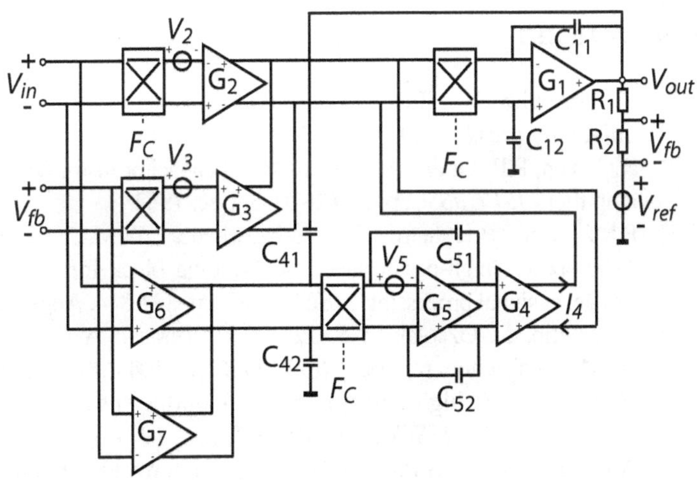

Fig. 4-17 *Chopper offset-stabilized chopper current-feedback instrumentation amplifier.*

4.2.6 Chopper offset-stabilized chopper instrumentation amplifier

A chopper offset-stabilized chopper current-feedback instrumentation amplifier is presented in figure 4-17. Since the chopped integrator acts as a band-pass filter for the input signal G_6 and G_7 are effectively AC coupled, therefore, the low frequency gain is determined by G_2 and G_3.This topology has also been extended to an iterative offset-stabilized instrumentation amplifier [4.18]. An instrumentation amplifier like this has the potential of achieving a very low offset, since every source of offset has been offset-stabilized. However it requires 24 gain stages.

4.3 Conclusions

In this chapter the topologies of operational amplifiers discussed in the previous chapter have been extended to include current-feedback instrumentation amplifiers. Of these topologies, the chopper offset-stabilized instrumentation amplifier offers the best option for obtaining a low-noise instrumentation amplifier.

4.4 References

[4.1] P. Horowitz, W. Hill, "*The arts of electronics*", Cambridge university press, 1989.

[4.2] R.C. Yen, P.R. Gray, "A MOS switched-capacitor instrumentation amplifier", *IEEE JSSC*, pp. 1008–1013, Dec. 1982.

[4.3] P.M. van Peteghem, I. Verbauwhede, W.M.C. Sansen, "Micropower high-performance SC building block for integrated low-level signal processing", *IEEE JSSC*, pp. 837–844, Aug. 1985.

[4.4] K. Martin, L. Ozcolak, Y.S. Lee, G.C. Temes, "A differential switched-capacitor amplifier", *IEEE JSSC*, pp. 104–106, Feb. 1987.

[4.5] H. Krabbe, "A high-performance monolithic instrumentation amplifier", *IEEE ISSCC*, pp. 186–187, Feb. 1971.

[4.6] A.P. Brokaw, M.P. Timko, "An improved monolithic instrumentation amplifier", *IEEE JSSC*, pp. 417–423, Dec. 1975.

[4.7] R.J. van de Plassche, "A wide-band operational amplifier with a new output stage and a simple frequency compensation", *IEEE JSSC*, pp. 347–352, Dec. 1971.

[4.8] R.J. van de Plassche, "A wide-band monolithic instrumentation amplifier", *IEEE JSSC*, pp. 424–431, Dec. 1975.

[4.9] J.H. Huijsing, "*Operational amplifiers theory and design*", Dordrecht: Kluwer, 2001.

[4.10] B.J. van den Dool, J.H. Huijsing, "Indirect current feedback instrumentation amplifier with a common-mode input range that includes the negative rail", *IEEE JSSC*, pp. 743–749, July 1993.

[4.11] J.H. Huijsing, B.Shahi, "Accurate voltage-to-current converters for rail-sensing current-feedback instrumentation amplifiers", US patent Nr. 7,202,738, Apr. 10, 2007.

[4.12] G.H. Hamstra, A. Peper, C.A. Grimbergen, "Low-power low-noise instrumentation amplifier for physiological signals", *Medical & Biological Eng. & Comput.*, pp. 272–274, May 1984.

[4.13] M.S.J. Steyaert, W.M.C. Sansen, "A micropower low-noise monolithic instrumentation amplifier for medical purposes," *IEEE JSSC*, pp. 1163–1168, Dec. 1987.

[4.14] R.F. Yazicioglu, P. Merken, R. Puers, C. Van Hoof, "A 60μW 60nV/√Hz readout front-end for portable biopotential acquisition systems", *IEEE JSSC*, pp. 1100–1110, May 2007.

[4.15] M.A.P. Pertijs, W.J. Kindt, "A 140dB-CMRR current-feedback instrumentation amplifier employing ping-pong auto-zeroing and chopping", IEEE ISSCC, pp. 324–325, Feb. 2009.

[4.16] J.H. Huijsing, "Instrumentation amplifiers developments", *AACD workshop*, Pavia, Apr. 2008.

[4.17] A.T.K. Tang, "A 3 μV-offset operational amplifier with 20nV/√Hz input noise PSD at DC employing both chopping and autozeroing", *IEEE ISSCC*, pp. 386–387, Feb. 2002.

[4.18] J.H. Huijsing, B.Shahi, "Chopper chopper-stabilized instrumentation and operational amplifiers", US patent Nr. 7,132,883, Nov. 7, 2006.

class-AB biased operational amplifier composed of G_2 and G_1. The stabilization loop around G_3 consists of a chopper amplifier, an integrator, a sample-and-hold, and G_4, which serves as an auxiliary input of G_3. The chopper amplifier consists of CH_2, G_6 and CH_1, while the integrator is composed of G_5 and capacitors C_{51} and C_{52}. It should be noted that all the choppers operate at the same clock frequency. The intermediate stage G_2 stage is mainly used to simplify the design. The current summing and class-AB biasing do not have to be combined in this way. The two stages can be combined, as will be demonstrated in the implementation described in section 5.3.

Due to feedback, the offset V_{os} appears at the input of chopper CH_2. As shown in figure 5-1, the amplifier is configured in an inverting topology via resistors R_1 and R_2, but the system works with any negative feedback network. The modulated offset voltage is converted into a current by G_6 and then demodulated by CH_1. The resulting DC current, which is proportional to the input offset voltage, is integrated. The integrator then applies an offset compensating voltage at the input of G_4. This transconductance acts as an auxiliary input for G_3 and applies an offset compensating current to the output of G_3. For as long as there is offset present at the input of the entire amplifier, the integrator will try to reduce it. Without taking charge injection effects into account, the residual offset due to finite gain can then be expressed by:

$$V_{os,res,gain} = \frac{A_{03} V_{os}}{A_{04} A_{05} A_{06}}, \tag{5-1}$$

where V_{os} is the uncompensated offset of the main amplifier consisting of G_3, G_2 and G_1, and A_{0i}, i=3, 4, 5, 6, are the DC voltage gains of the appropriate amplifier stages. This means that in order to reduce the offset of the main amplifier from, say, 10 mV to 0.1 μV, the gain of the stabilization loop should be at least 100 dB higher than the gain of G_3.

The circuit shown in figure 5-1 can be considered as a multi-path amplifier in which the cascaded transconductances G_3, G_2 and G_1 form the low-gain/high-frequency path, while the cascaded transconductances G_6, G_5, G_4, G_2, and G_1 form the high-gain/low-frequency path. Capacitors C_{11}, C_{12}, C_{21}, and C_{22} implement nested Miller frequency compensation [5.4] and are chosen to achieve a 60-degree phase margin for a 1 MHz unity gain frequency with a 100 pF load. The two capacitors C_{41} and C_{42}, together with the

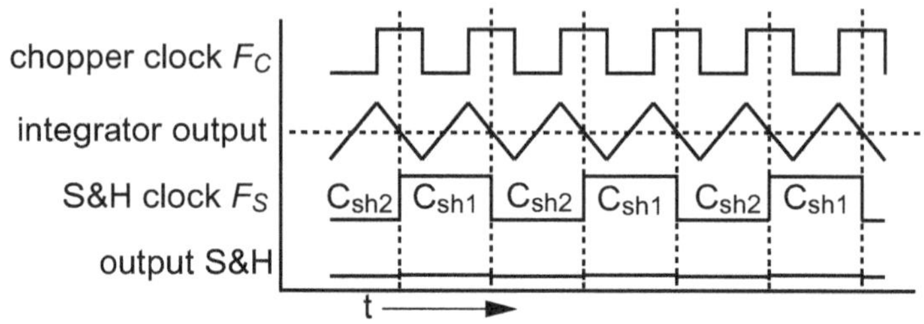

Fig. 5-2 *Timing diagram. It is indicated if C_{sh1} or C_{sh2} is sampling the output of G_5.*

integrator capacitors C_{51} and C_{52}, form a hybrid-nested Miller frequency compensation, as discussed in section 3.3.2.

Chopper ripple reduction

The modulated offset of G_6 gives rise to chopper ripple, in the form of a triangular wave at the output of G_5, which in turn gives rise to a triangular wave at the input of the whole amplifier. The input referred peak-to-peak voltage of this triangular wave can be approximated by:

$$V_{in-pp} = \frac{V_6 G_6 G_4}{2F_C C_5 G_3}, \tag{5-2}$$

where V_6 is the offset of G_6, F_C is the chopper frequency, and C_5 is the integrator capacitance. For use as a general-purpose operational amplifier, the amplitude of this ripple should be reduced below the noise floor.

From equation (5-2) it can be seen that the amplitude of the ripple can be minimized by decreasing the transconductance G_6 and G_4, and increasing the size of the integrating capacitors C_{51} and C_{52}. However, increasing the capacitance size increases the chip area. While decreasing the transconductance of G_6 increases the input-referred noise of the amplifier. Alternatively, the chopping frequency can be increased, although this causes more residual offset due to the effect of charge injection in the chopper switches. It is also possible to auto-zero G_6, which is done in another implementation discussed in section 5.3.

In this design the chosen solution is to implement a sample-and-hold circuit after the integrator, as shown in figure 5-1. The circuit samples the triangular wave at the integrator output, thereby eliminating the ripple. The timing of the sample moment is not critical, although the dynamic range of the integrator is used optimally when the triangular wave is sampled at the zero crossings. To achieve a quasi-continuous output signal, two sample-and-hold circuits have been used. Their timing is arranged in such a way that while one sample-and-hold capacitor C_{sh1} is sampling, the other sample-and-hold capacitor C_{sh2} is driving G_4. The timing diagram of the sample-and-hold is shown in figure 5-2.

The offset of G_5 also contributes to the residual offset. This is because it appears as a chopped voltage that charges and discharges the parasitic output capacitance C_{p6} of G_6. The required current is provided by G_6, which means that a voltage must be present at its input. Hence there will be a residual offset at the input of the amplifier. This residual offset is given by:

$$V_{os,res,par} = \frac{4V_5 F_C C_{p6}}{G_6}. \tag{5-3}$$

To minimize this offset, the transconductance of G_6 should be maximized, while its parasitic capacitance should be minimized. However, as can be seen from equation (5-2), maximizing G_6 will increase the ripple. A designer could minimize the parasitic output capacitance of G_6 by using minimum-sized transistors and by minimizing the length of the metal wires connecting G_6 to the chopper. A compromise will need to be made between residual offset and ripple amplitude. Of course, the ripple amplitude at the output of the integrator is bounded by the power supply voltage of the integrator, since the integrator cannot be allowed to clip.

A low transconductance of 4 μA/V is implemented to measure this amplifier with clock frequencies down to 1 kHz. A clock frequency of 1 kHz combined with integrator capacitors of 32 pF, a 15 mV worst case offset, and a G_6 of 4 μA/V would lead to a triangular wave at the output of the integrator with a peak-to-peak value of 1 V. With lower clock frequencies the integrator would saturate.

Another possibility which increases the complexity of the circuit, but does not have conflicting trade-offs, is to cancel the integrator's offset. In order to do this, a nested offset-stabilization loop is implemented around the integrator. The complete topology can be seen in figure 5-3.

The offset V_5 of G_5 appears at the input of chopper CH_1 as a square wave. This signal is converted into a current and then demodulated by the sense amplifier G_9 and chopper CH_3, respectively. An integrator composed of

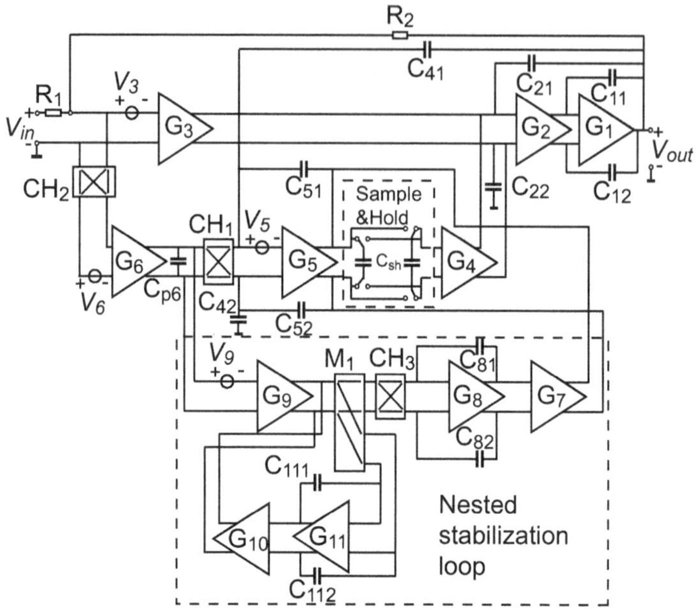

Fig. 5-3 *Chopper offset-stabilization using an active integrator with a sample-and-hold to filter the chopper ripple and a nested integrator offset-stabilization loop.*

G_8, C_{81}, and C_{82}, integrates this current and then drives an offset compensating current into the output of G_5 via its auxiliary input G_7. All the choppers work at the same clock frequency F_C.

The offset V_9 of G_9 also needs to be reduced because it is a source of additional residual offset and output ripple. For this reason the multiplexer M_1, integrator G_{11}, and auxiliary input G_{10} are added, which auto-zero G_9 at half the chopper clock frequency. This technique is similar to the iterative offset-stabilization approach discussed in section 3.4.1.

To further investigate the behaviour of this topology, several parts of the circuit can be bypassed. By doing this, the three-stage amplifier consisting of G_1, G_2 and G_3 can be measured. The offset-stabilization loop and nested offset-stabilization loop can be turned on separately, while the sample-and-hold can be switched on or off.

5.2.2 Circuits

So far only the topology has been discussed. This section focuses on the circuits with which this design has been implemented. First, the output stage is discussed. Secondly, the input stages are discussed. Both input stages are designed with the same topology, however, they are biased at different current levels. After that the integrator is discussed. Lastly, the biasing technique and bias current generator is discussed. The values of the transistor width W and length L have been omitted because they were not chosen optimally. In the implementations discussed in section 5.3.2, the values of W and L will be shown.

Output stage

In the implementation of the topology shown in figure 5-3, a class-AB output stage was chosen for stages G_1 and G_2. This topology is shown in figure 5-4. The quiescent current through the output transistors M_1 and M_2 is controlled by two trans-linear loops designed with M_{1-4} and M_{5-8} [5.5]. The general idea is that the gate source voltages of M_1 and M_5 are controlled by these translinear loops such that:

$$V_{gs1} + V_{gs4} = V_{gs3} + V_{gs2}\,. \tag{5-4}$$

When M_3 and M_4 operate at the same drain current density, their gate-source voltages are equal, so the gate-source voltages of M_1 and M_2 should also be equal. This means that the current through M_1 when $I_{out} = 0$, i.e. the quiescent current, is equal to:

$$I_q = \frac{W_1 L_2}{W_2 L_1} I_5\,. \tag{5-5}$$

In this design the output transconductance was designed to be 2500 μA/V in order to drive a 100 pF output capacitor with a unity gain frequency of 4 MHz. Since this stage has two preceding stages, and Miller compensation is used, the overall amplifier has a unity gain of 1MHz [5.4]. Since M_1 and M_5 were only operating in moderate inversion, the quiescent current I_q through M_1 and M_5 is 250 μA.

Transistors M_9 and M_{10} are a copy of M_4 and M_8. They are added to improve the offset voltage of the operational amplifier, by keeping the drain source voltage of cascode transistors M_{12} and M_{14} almost equal. Without the transistors M_9 and M_{10} the drain-source voltages of M_{12} and M_{14} would differ by a few volts, which in turn would cause their gate-source voltages to differ, thus causing an input offset.

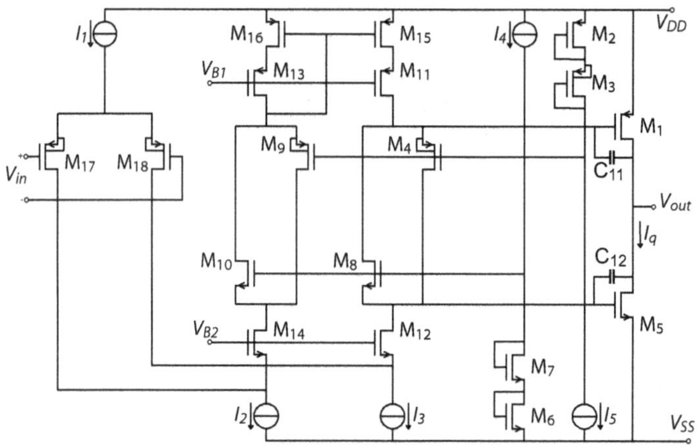

Fig. 5-4 *Two-stage class-AB output operational amplifier implementation of G_1 and G_2.*

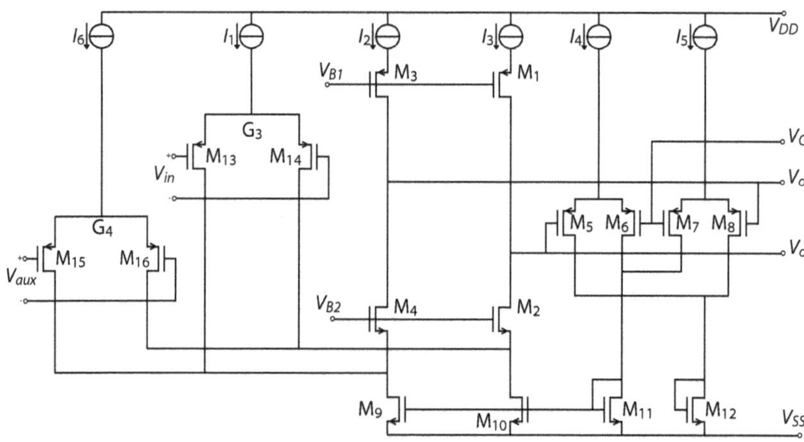

Fig. 5-5 *Folded cascode with auxiliary input for offset compensation and a common-mode feedback circuit.*

Input stages

The input stages G_3 and G_6 are implemented using a folded cascode topology. The circuit of G_3 with its auxiliary input G_4 can be seen in figure 5-5. PMOS input transistors M_{13} and M_{14} are used to implement G_3. Cascode transistors M_{1-4}, current sources I_2 and I_3, and current sources M_9 and M_{10} are used to implement a fully-differential folded cascode operational amplifier. The current through M_9 and M_{10} is controlled by a common-mode feedback circuit designed with a differential-difference amplifier composed of long PMOS transistors M_{5-8}. These transistors are long to achieve a low transconductance and a wider linearity range. This amplifier compares the output common-mode voltage to the voltage V_{CM} and adjusts the current through M_{11} to keep the output common-mode voltage equal to V_{CM}.

M_{15} and M_{16} form the auxiliary input for the offset compensating voltage. Its output current is summed together with the output currents of M_{13} and M_{14}. A MOS transistor biased with a drain current I_d in weak inversion has the transconductance $g_m \approx I_d / nV_{th}$. In the process used, nV_{th}=50mV at room temperature. Thus, the differential input V_{aux} is effectively limited to about 50 mV. The transconductance of G_4 is chosen a factor of four lower than the transconductance of G_3 by making I_6 four times smaller than I_1. This means that the $\sqrt{kT/C}$ noise voltage associated with the sample-and-hold is reduced only by a factor of four, when referred to the input. Probably a better implementation would involve degenerating the differential input pair with a resistor R_{deg}. A transconductance of $g_m \approx 1/R_{deg}$ would then be obtained and the range of V_{aux} is given by $I_d R_{deg}$, which can be made much larger then 50 mV. In the implementations discussed in section 5.3 and chapter 6, the auxiliary amplifier was implemented in this way. This wider limit would make better use of the integrator's output dynamic range.

For G_6, the same topology is used, but without the auxiliary input. From equation (5-2) it can be seen that the ripple amplitude is proportional to G_6 and inversely proportional to the chopper frequency F_c. Therefore, the transconductance of G_6 is designed to be 100 times lower than that of G_3 by applying bias currents that are 100 times lower. This choice was made in order to contain the ripple at the output of the integrator at low chopper frequencies and to be able to measure at a 1 kHz chopper frequency. The low transconductance actually improves the DC voltage gain of G_6 with respect to G_3. However, it destroys the low frequency noise performance of the whole amplifier, since this is dominated by the white noise of G_6.

Integrator

In order to achieve less then 0.1 μV offset, while the initial uncompensated offset is 10 mV, the DC voltage gain of G_4, G_5 and G_6 has to be at least 100 dB more than of G_3, as can be seen in equation (5-1). The DC gain of G_4 is 4 times less than that of G_3. The DC gain of G_6 is about 50 dB over corners. A voltage divider with a voltage ratio of 25 was added after the integrator, to make use of the dynamic range of the integrator. These decisions led to required 80dB DC voltage gain.

Therefore, the operational amplifier G_5, which is used as an integrator, was implemented by a two-stage amplifier as shown in figure 5-6. Transistors M_3 and M_4 form an input differential pair, while transistors M_1 and M_2 form the output gain stage. Miller capacitances C_{M1} and C_{M2} are used to assure local stability. A resistor $R_{1\text{–}4}$ network is used to sense the output common-mode voltage. This resistor divider also divides the output of the integrator capacitances by a factor of 25. Transistors M_5 and M_6 make the output common-mode voltage equal to a set voltage V_{CM} at the source of the replica transistor M_7. With this common-mode feedback, a rail-to-rail output can be created, which is needed to handle the chopper ripple. To make optimal use of the output dynamic range, V_{CM} could be fixed at $0.5V_{DD}$.

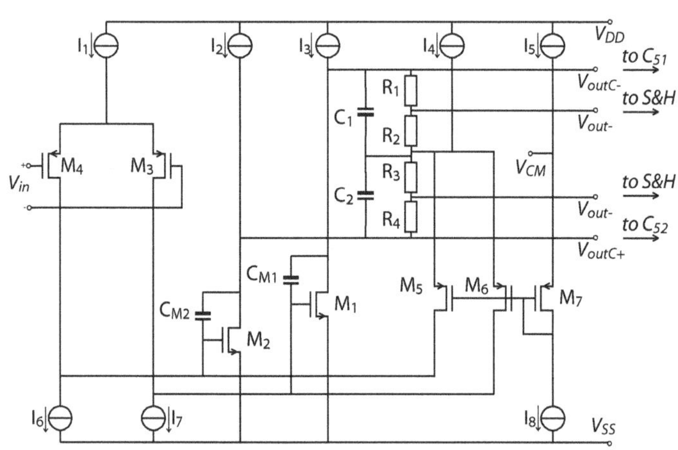

Fig. 5-6 *Two-stage class-A operational amplifier with resistive common-mode feedback.*

Bias generator

All the transconductance setting transistors in the amplification stages of the amplifier are of the PMOS type. To implement a constant transconductance over temperature, a bias generator (illustrated in figure 5-7) was used. This biasing is used in all implementations. In this part of the book it will be shown that this biasing technique ensures a constant transconductance over temperature.

In weak inversion, the effective gate-source voltage of a PMOS transistor is given by:

$$-V_{gseff} = V_T - V_{gs} = n_p V_{th} \ln\frac{I_d}{I_s}, \qquad (5\text{-}6)$$

where n_p is the weak inversion slope factor of a PMOS transistor, which is about 2 for the used CMOS processes, and I_s is the specific current, which for a PMOS transistor is given by:

$$I_s = 2n\mu_p C_{ox} V_{th}^2 \frac{W}{L}. \qquad (5\text{-}7)$$

So when M_1 is x times wider then M_2 the current through R_{ptim} is:

$$I_{ptim} = \frac{n_p V_{th} \ln x}{R_{ptim}}, \qquad (5\text{-}8)$$

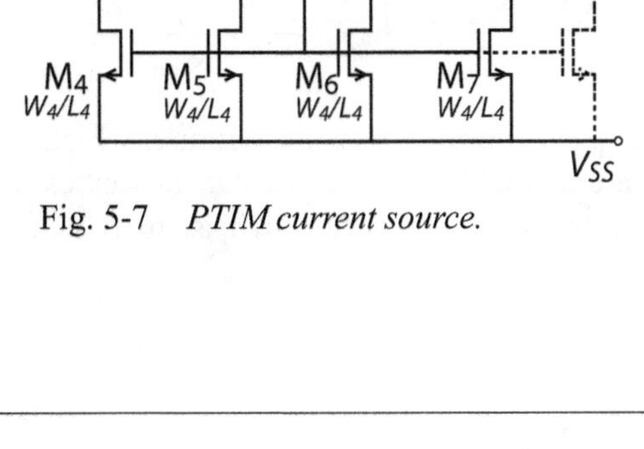

Fig. 5-7 *PTIM current source.*

which is actually a current proportional to absolute temperature. When a PMOS transistor operating in weak inversion is biased with a multiple of this current, its transconductance will be proportional to the value of the resistor since:

$$G_m = \frac{I_d}{n_p V_{th}} \propto \frac{1}{\mathrm{R_{ptim}}}. \tag{5-9}$$

For strong inversion the effective gate-source voltage of a PMOS transistor is given by:

$$-V_{gseff} = V_T - V_{gs} = \sqrt{\frac{2}{\mu_p C_{ox}} \frac{L}{W} I_d}. \tag{5-10}$$

The voltage over the resistor $\mathrm{R_{ptim}}$ can be expressed as:

$$V_R = \sqrt{\frac{2}{\mu_p C_{ox}} \frac{L_2}{W_2} I_d} - V_{T2} - \sqrt{\frac{2}{\mu_p C_{ox}} \frac{L_1}{W_1} I_d} + V_{T1} \approx \left(\sqrt{\frac{L_2}{W_2}} - \sqrt{\frac{L_1}{W_1}}\right)\sqrt{\frac{2}{\mu_p C_{ox}} I_d}. \tag{5-11}$$

Under the condition that the threshold voltages of M_1 and M_2 are equal and substituting $V_R = R_{ptim} I_d$ $L_1 = L_2$ and $W_1 = x W_2$ this leads to:

$$I_d = \frac{1}{\mathrm{R}^2_{\mathrm{ptim}}} \frac{2}{\mu_p C_{ox}} \frac{L_1}{W_1} \frac{(1-\sqrt{x})^2}{x}. \tag{5-12}$$

This current is proportional to the inverse of the mobility μ_p, hence the name PTIM. The mobility is proportional to $\mathrm{T}^{-\mathrm{n}}$, where $\mathrm{n} \approx 2.2$ [5.6]. The value of n depends on doping concentration. If a multiplication of this current is used to bias a PMOS transistor in strong inversion, it can be seen that, again, its transconductance is proportional to the resistor value since:

$$G_m = \sqrt{\mu_p C_{ox} \frac{W}{L} I_d} \propto \frac{1}{\mathrm{R_{ptim}}}. \tag{5-13}$$

Assuming that $\mathrm{R_{ptim}}$ is constant over temperature, which is true for thin-film resistors, which are available in many CMOS processes, it is clear that this biasing leads to a transconductance, which is, to first order, constant over temperature. However, both the transconductance setting transistor and the bias current setting transistor should operate in the same region. In other words, this technique works best if the current density of M_2 matches the current densities of the transconductance setting transistors in the amplification stages. The transistors should also be of the same type to avoid influences of the p and n type doping.

Current consumption

In simulations the amplifier draws a total current of 770 μA. In table 5-1, both the value of the transconductance and the current consumption of each stage as well as the values of the capacitors are given. Although this design isn't designed for optimal current consumption, it can be seen that the nested stabilization loop that compensates for the offset of the integrator G_5 consumes only 12% of the total power.

Table 5-1. *Capacitor values, transconductance, and current consumption of each stage.*

Stage	Transcon-ductance	Current consumption	Capacitors	Cap value
G_1	2.5 mA/V	260 μA	C_{11}+C_{12}	8 pF
G_2	130 μA/V	110 μA	C_{21} & C_{22}	60 pF
G_3	400 μA/V	180 μA		
G_4	100 μA/V	12.5 μA	C_{41} & C_{42}	1 pF
G_5	40 μA/V	40 μA	C_{51} & C_{52}	40 pF
G_6	4 μA/V	2 μA		
G_7	10 μA/V	1 μA		
G_8	40 μA/V	40 μA	C_{81} & C_{82}	10 pF
G_9	4 μA/V	2 μA		
G_{10}	1 μA/V	100 nA		
G_{11}	40 μA/V	40 μA	C_{111} & C_{112}	10 pF
Bias		100 μA		

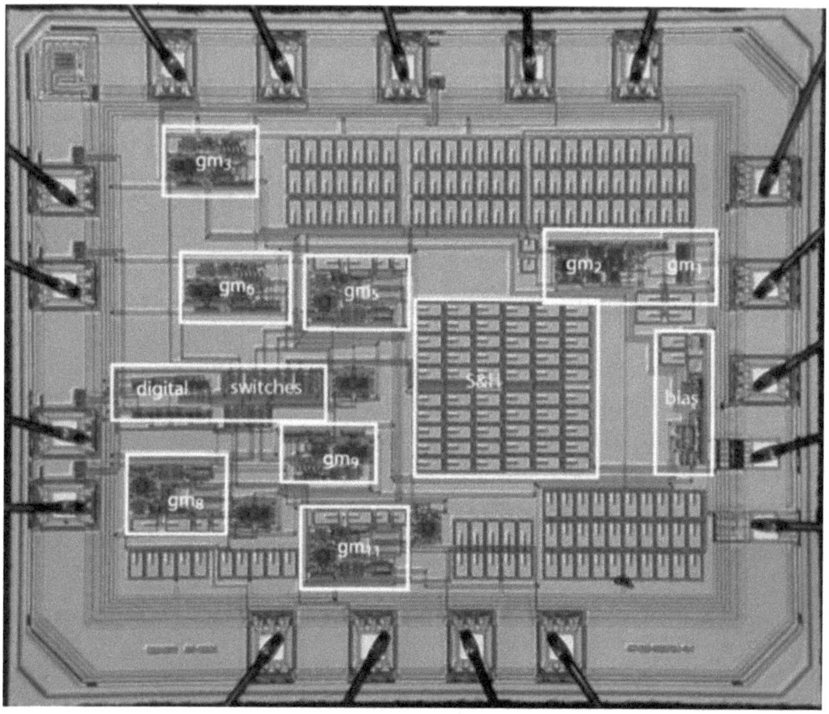

Fig. 5-8 *Chip micrograph.*

5.2.3 Measurement results

The design was implemented in a 0.7 μm CMOS process, with a chip area of 3.6 mm^2. The chip micrograph is shown in figure 5-8. The measured supply current is 700 μA from a 5 V supply voltage.

Frequency response

In figure 5-9 the measured frequency response of the complete amplifier is shown. With a 50 pF load, the unity-gain frequency is 1.3 MHz with a 56-degree phase margin. The circuit is also stable with a closed-loop DC gain of 80 dB and has a 20 dB per decade roll off, which demonstrates the efficacy of the multi-path hybrid-nested Miller compensation.

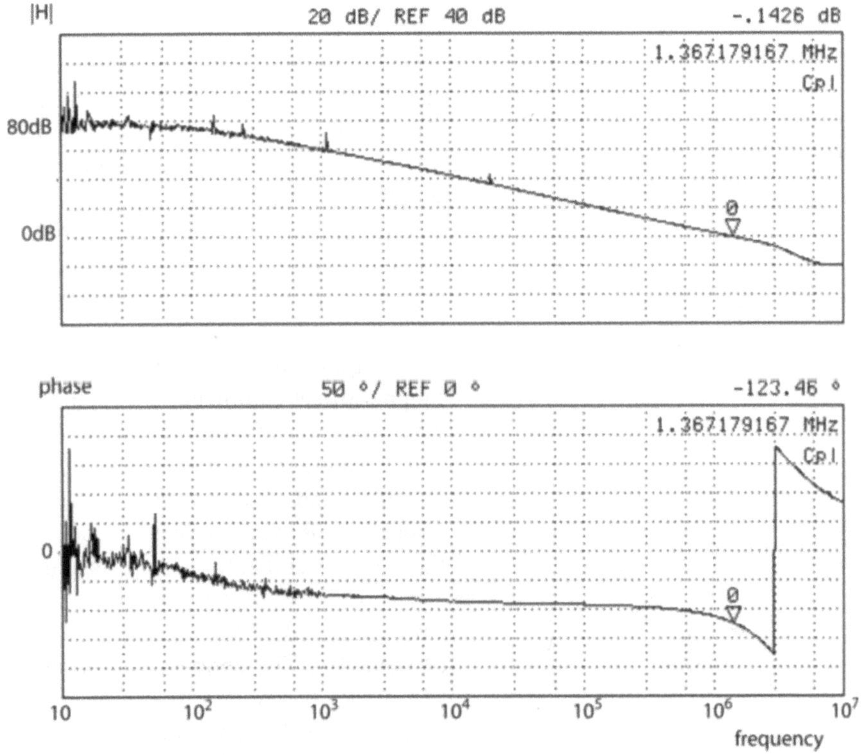

Fig. 5-9 *Bode plot at a DC gain of 80 dB with a 50 pF load.*

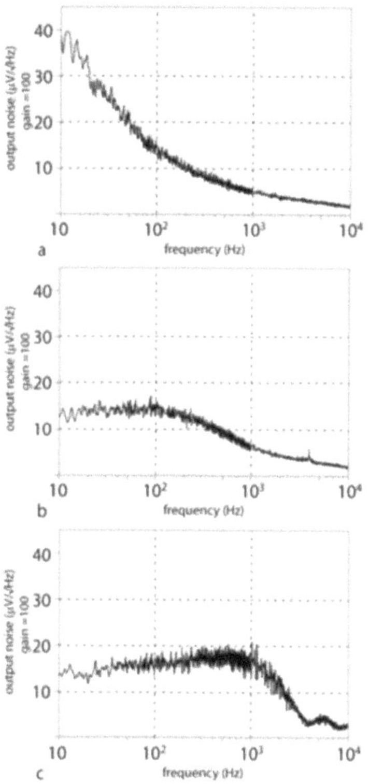

Fig. 5-10 *Spectrum of the output noise (gain 40 dB) of: (a) the amplifier without offset-stabilization; (b) the amplifier with chopper (4 kHz) offset-stabilized turned on; (c) as b with the sample-and-hold (2 kHz) and nested stabilization loop turned on.*

Noise spectrum

The noise spectrum is measured at the output of the amplifier. For these measurements, the amplifier is used in an inverting configuration with a DC gain of 40 dB. The –3dB bandwidth is around 10 kHz. The output noise spectrum of the amplifier with the offset stabilization technique turned off is shown in figure 5-10a. Low frequency $1/f$ noise can be seen.

The output noise spectrum with the chopper offset-stabilization turned on is shown in figure 5-10b. The $1/f$ noise is now suppressed by the chopper amplifier, and an increased white-noise level at low frequencies can be observed since the stabilization loop produces more noise than the main amplifier. The input referred noise density is of 140 nV/√Hz. The chopper frequency is 4 kHz and the chopper ripple is visible as a small peak. When observed on an oscilloscope, the ripple is triangular with an amplitude of about 250 μV amplitude ripple.

In figure 5-10c, the output noise spectrum is shown with the sample-and-hold and nested stabilization loop turned on. In this measurement the sample-and-hold runs at 2 kHz, and it can be seen that the chopper ripple is now effectively suppressed. When observed on an oscilloscope, the ripple is below the noise level. However, since the sample-and-hold under-samples the wide band noise of the stabilization loop, the total noise increases somewhat.

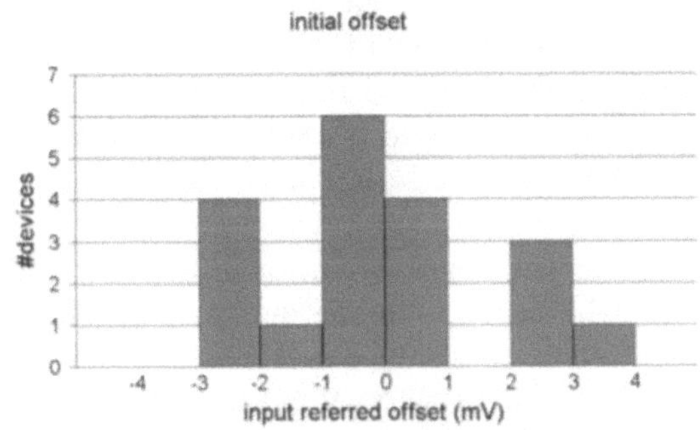

Fig. 5-11 *Initial offset without any stabilization.*

Offset

First, the initial offset of the amplifier was measured without activating any offset-stabilization circuits. Secondly, the chopper offset-stabilization circuit was activated, resulting in the circuit of figure 5-1 without the sample-and-hold activated. Next, the offset stabilization loop of the integrator was activated in order to realise the circuit of figure 5-3, but without the sample-and-hold. Finally, the sample-and-hold was also enabled. The measurements were made on 19 samples using an external clock generator. The measurements results are summarized in table 5-2.

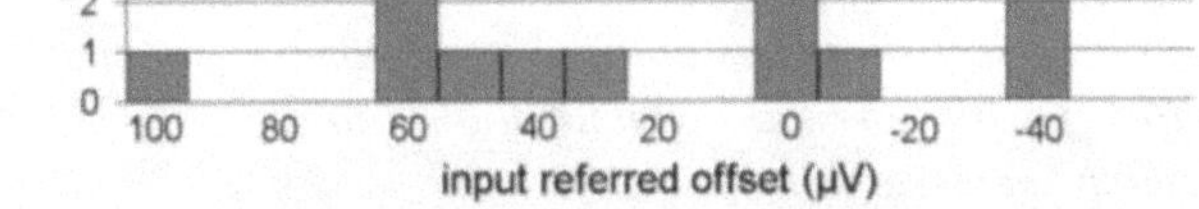

Fig. 5-12 *Offset of a chopper offset-stabilized operational amplifier without nested stabilization loop.*

First, the offset of the amplifier without offset-stabilization is presented. A histogram of all 19 samples is shown in figure 5-11. A 0.1mV average offset with a 1.5 mV average absolute deviation was measured. The average absolute deviation from the mean was used as a figure of merit for the spread in offset voltage, because the measured offset distribution was not truly Gaussian.

Second, the chopper offset-stabilization loop was activated, but with both the nested integrator offset-stabilization loop and the sample-and-hold switched off. A histogram of all 19 samples is shown in figure 5-12. With an externally applied clock frequency of 16 kHz, a 16 μV average offset with a 30 μV average absolute deviation was measured. Consequently, the spread is 50 times lower than without offset compensation. The main cause of residual offset is the offset of the integrator. This corresponds with the results of simulation.

Third, the nested offset-stabilization loop of the integrator was activated. A –1.4 μV average offset with a 0.5 μV average absolute deviation was measured. A histogram is presented in figure 5-13. This shows that this nested stabilization loop lowers the offset of the whole amplifier by more than a factor of 10 and reduces the spread by a factor of 60. The input-referred amplitude of the triangular ripple can be seen on an oscilloscope and can be as high as 250 μV.

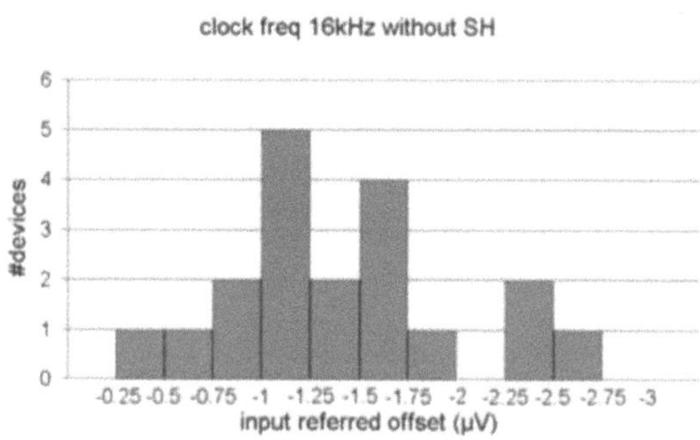

Fig. 5-13 *Offset of a chopper offset-stabilized operational amplifier with nested stabilization loop excluding sample-and-hold.*

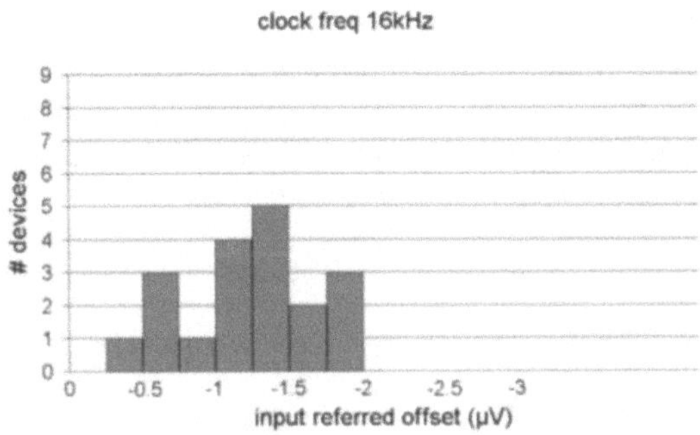

Fig. 5-14 *Offset of chopper offset-stabilized operational amplifier according to figure 5-3 including the sample-and-hold.*

Finally, the sample-and-hold was activated to reduce the chopper ripple below the noise level, and offset was measured at three different clock frequencies. Theoretically, the offset should not be influenced by the sample-and-hold action. However, due to the reduction of the chopper ripple, some on-chip capacitive cross-talk effects were also reduced, which resulted in a somewhat lower average offset and average absolute deviation. The histogram of the measurements with a 16 kHz clock is shown in figure 5-14.

Table 5-2. *Offset measurement statistics.*

Circuit complexity	Clock frequency (kHz)	Average offset (μV)	Absolute average offset deviation (μV)
A initial amplifier	16	100	1,500
B figure 5-1 without S&H	16	16	30
C figure 5-3 without S&H	16	−1.4	0.5
D figure 5-3	16	−1.22	0.33
	4	−0.8	0.13
	32	−1.81	0.67

With a 4 kHz clock frequency, an average offset plus average absolute deviation of less than 1 μV was achieved. From table 5-2 it can be seen that the offset spread increases with the clock frequency. Therefore, it can be concluded that clock-frequency-dependent capacitive cross-talk and charge injection are the main sources of the remaining residual offset.

5.3 Chopper and auto-zero offset-stabilized operational amplifier

In this section of the book, a chopped and auto-zeroed offset-stabilized operational amplifier will be presented. This design was implemented after the instrumentation amplifier presented in section 6.3, and in parallel with the instrumentation amplifier presented in section 6.2. This amplifier was implemented to investigate the performance of a chopper and auto-zero offset-stabilization topology, as discussed in chapter 3, and to reduce power consumption.

First the topology is presented. After that the transistor implementations of the transconductances used in this realization are shown. Then the measurement results are shown in section 5.3.3. This section ends with conclusions.

5.3.1 Topology

The topology presented in this section is based on a two-stage amplifier. Around the first stage, a chopped and auto-zero offset-stabilized loop is implemented, as can be seen in figure 5-15. Stages G_1 and G_2 form a main amplifier implemented as a two-stage class-AB amplifier. The stabilization loop around G_2 consists of a chopped auto-zeroed amplifier, and an integrator. Stage G_3 is an auxiliary input for this two-stage amplifier. The integrator, implemented with G_4 and capacitors C_{41} and C_{42}, applies an offset compensating voltage to the inputs of G_3. The chopped and auto-zeroed amplifier G_5 measures the offset of the main amplifier. Stages G_5 and G_4 are implemented as a folded cascode and a gain-boosted folded cascode, respectively. Capacitors C_{11-12}, C_{31-32} and C_{41-42} form the hybrid-nested Miller frequency compensation, as discussed in section 3.3.2.

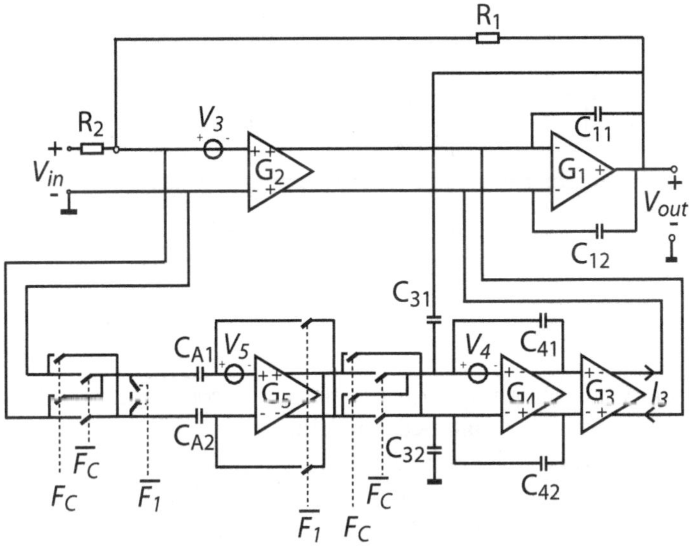

Fig. 5-15 *Chopped Auto-zeroed offset-stabilized operational amplifier.*

Instead of sampling the integrator output voltage, as was discussed in section 5.2.1, in this design the auto-zero technique is used to reduce the chopper ripple. As shown in equation (5-2), the chopper ripple is proportional to the initial offset of the chopped amplifier. Therefore, auto-zeroing this offset significantly reduces the chopper ripple, as discussed in chapter 3.3.5. Clock $\overline{F_1}$ is used to drive the auto-zero action.

In this design, the chopper switches, driven by clock F_C, are also used as auto-zero switches, i.e. to implement clock F_1. Therefore, they are drawn as separate switches in figure 5-15. All clocks are generated synchronously with an on-chip oscillator and logic. The timing diagram is shown in figure 5-16.

The chopper clocks F_c and $\overline{F}_c$ alternate in time in order to drive the input more symmetrically. Because logic is already needed to implement clock F_C, this alternating algorithm only adds an insignificant portion to the complexity.

Comparing this design with the previous design, the input stage G_5 has a higher transconductance. Therefore, the white noise level should be much lower. However, the switches driven by F_c and $\overline{F}_c$ can limit the white noise

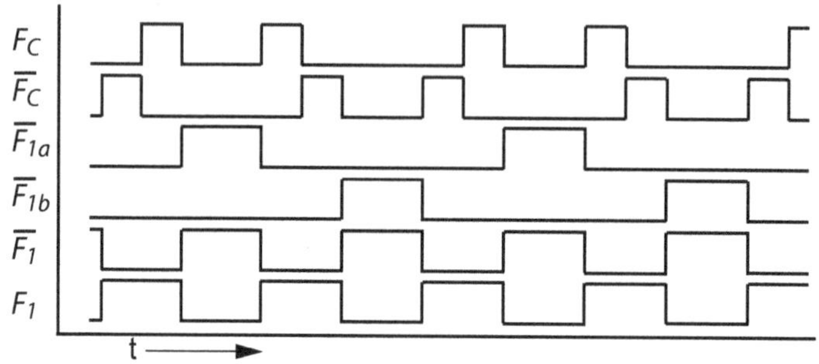

Fig. 5-16 *Timing diagram.*

floor of the whole amplifier, because the noise contribution of the R_{on} of minimum size switches is significant in comparison to the amplifier G_5. In this case, these switches are chosen with W/L=11/0.7 μm instead of minimum size switches of W/L=2/0.7 μm, which means that their charge injection is a factor 5.5 more significant. This in turn leads to a higher residual offset.

The offset of G_4 influences the residual offset, because it appears as a chopped voltage that charges and discharges the parasitic output capacitance C_{p5} of G_5. The required current is provided by G_5, which means that a voltage must be present at its input. Hence there is a residual offset at the input of the amplifier. This residual offset is given by:

$$V_{ores} = \frac{4V_4F_CC_{p5}}{G_5}. \qquad (5\text{-}14)$$

If C_{p5}=1 pF, F_C=30 kHz, and G_5=200 μA/V, then a 10 mV offset V_4 would cause a residual offset of 6μV. To be able to achieve a lower offset, an auto-zero offset-stabilization technique is used around the integrator, as shown in figure 5-17. The transconductance G_7 is auto-zeroed with clock $\overline{F_1}$. During clock F_1 the auto-zeroed transconductance G_7 measures the offset of integrator G_4, and applies current to capacitor C_{A5}, which functions as a passive integrator. Transconductance G_7 is implemented as a copy of G_5. Stage G_6 is an auxiliary input of G_4. The offset V_4 of integrator transconductance G_4 can be reduced down to:

$$\frac{V_{4res}}{V_4} = \frac{G_4}{A_{o7}G_6}, \qquad (5\text{-}15)$$

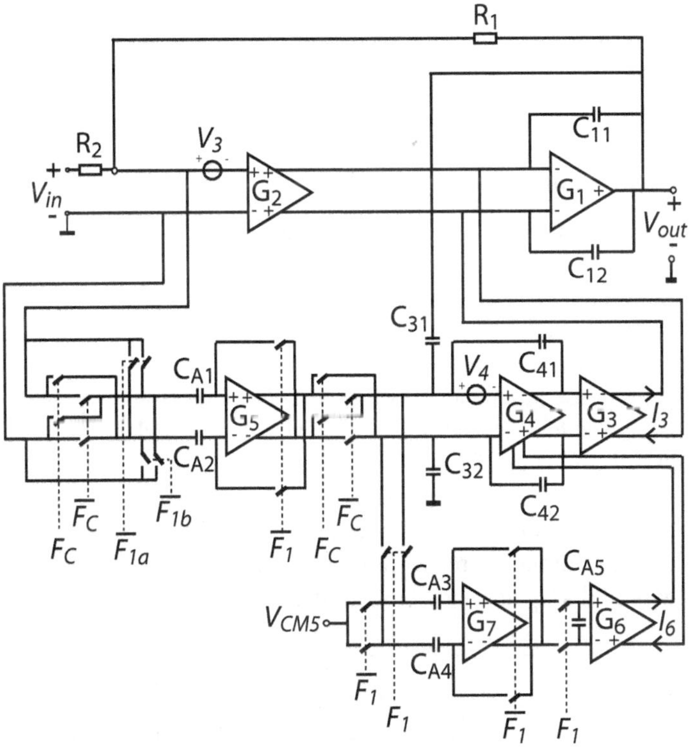

Fig. 5-17 *Chopped Auto-zeroed offset-stabilized operational amplifier with auto-zero offset-stabilized integrator.*

where A_{07} is the DC voltage gain of G_7. Transconductances G_4 and G_6 share the same output impedance.

Furthermore, during the auto-zero action the inputs of G_5 are alternatingly connected to the positive or negative input of the whole amplifier. Without doing this the charge injection would only be applied to one input, leading to a systematic cause for residual offset. In this implementation only the charge injection mismatch between the switches that are driven by $\overline{F}_{1a}$, $\overline{F}_{1b}$, F_c, and $\overline{F}_c$ introduces residual offset.

5.3.2 Circuits

So far the topology has been discussed. This part of the book focuses on the circuits with which this design has been implemented. First the implementation of G_1, G_2 and G_3 will be discussed. Then the implementation of the chopper amplifier G_5 is presented. Finally, the integrator amplifier G_4 is discussed. The transconductances are biased with the same biasing technique as described in section 5.2.2.

Input and output stage

In figure 5-18 the implementation of G_1, G_2 and G_3 is shown. In comparison to the realization shown in figure 5-4, this output stage G_1 consumes two times less quiescent current. Since a two-stage topology is used, only half the transconductance is needed to obtain the same unity gain frequency as a Miller compensated three-stage amplifier.

To improve the independence of quiescent current to the power supply voltage V_{DD}, cascode transistors M_9 and M_{10} are added to the class-AB biasing transistors M_4 and M_8. The quiescent current is dependent on the drain source voltage of M_4 and M_8, which is dependent on V_{DD}. The cascodes M_9 and M_{10} fix the drain source voltages of M_4 and M_8.

To keep the drain source voltage of M_{16} at the same level as M_{15}, transistors M_{17} and M_{18} are added. Since their drain current is twice as high

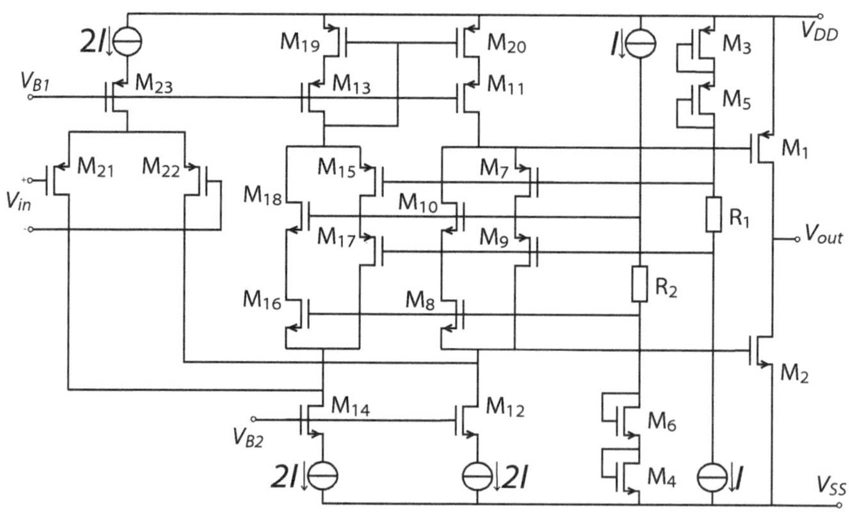

Fig. 5-18 *Two-stage class-AB operational amplifier with improved dependency of quiescent current to V_{DD} and degenerated auxiliary input.*

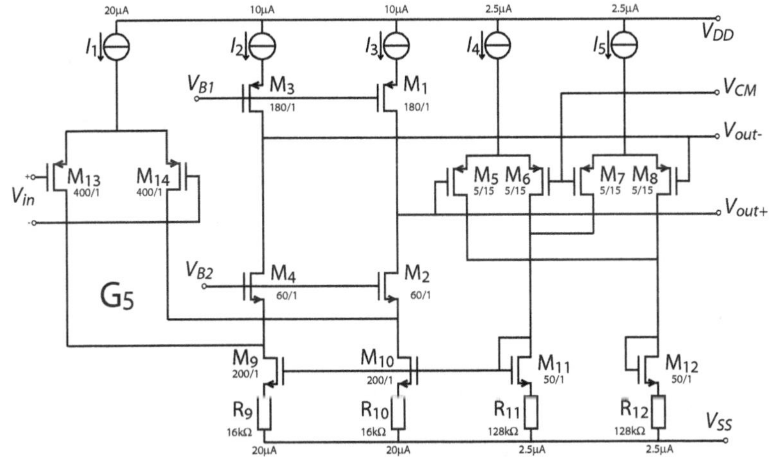

Fig. 5-19 *Folded cascode operational amplifier.*

as of M_8 and M_{10} at quiescent, their width is doubled to keep the same current density.

The auxiliary input G_3 is implemented as a degenerated differential pair. This degeneration makes the differential input range approximately 1 V, making the resistor divider in front of the auxiliary input, as used in the previous section, superfluous.

The chopper amplifier G_5 is implemented as a folded cascode amplifier, as shown in figure 5-19. Again, the common-mode feedback circuit is designed with a differential difference amplifier composed of long PMOS transistors $M_{5\text{–}8}$.

Integrator

In this design for the integrator operational amplifier G_4, the two-stage configuration was not used. Instead a single stage gain-boosted topology [5.7], as shown in figure 5-20, was used. The gain-boosting amplifiers are shown in figure 5-21.

Gain-boost amplifier G_p fixes the drain voltages of M_{13} and M_{14} to a voltage V_{B2}, while gain-boost amplifier G_n fixes the voltage over current sources I_2 and I_3 to V_{B1}, thus creating a very high output impedance. Again, the common-mode feedback circuit has been designed with a differential difference amplifier composed with long PMOS transistors M_{5-8}. In this topology the integrator does not have to be able to drive a large chopper ripple. Consequently, this common-mode feedback can be used because a rail-to-rail output stage is not necessary.

The gain-boost amplifiers G_p and G_n are depicted in figure 5-21. They both use input common-mode feedback to apply the voltages V_{B1} and V_{B2} to the inputs of the gain-boost amplifiers.

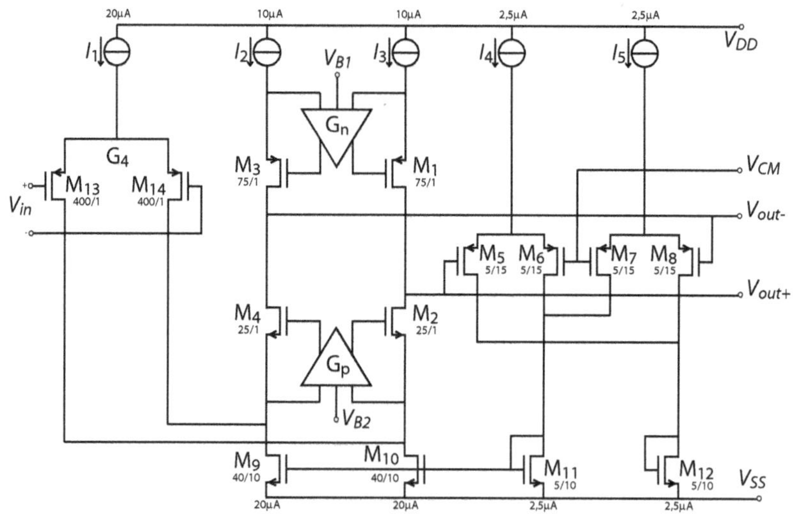

Fig. 5-20 *Gain-boosted folded cascode topology.*

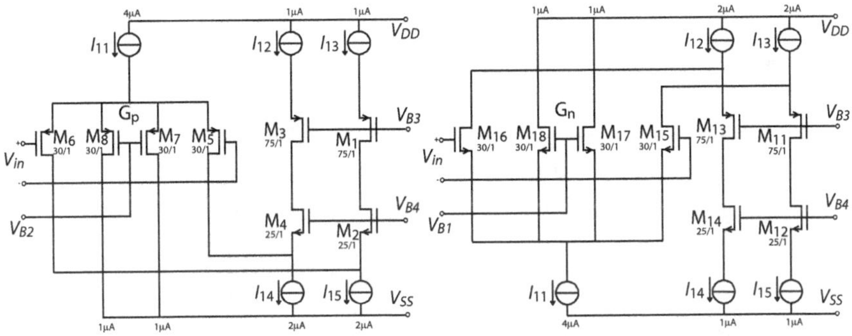

Fig. 5-21 *Folded cascodes amplifiers with PMOS and NMOS input pairs both using input common-mode voltage control.*

Table 5-3. *Capacitor values, transconductance, and current consumption of each stage.*

Stage	Transcon-ductance	Current consumption	Capacitors	Cap value
G_1	1 mA/V	55 μA	$C_{11}+C_{12}$	32 pF
G_2	200 μA/V	55 μA		
G_3	2.5 μA/V	7.5 μA	C_{31} & C_{32}	32 pF
G_4	200 μA/V	50 μA	C_{41} & C_{42}	24 pF
G_5	200 μA/V	55 μA	C_{A1} & C_{A2}	4 pF
G_6	50 μA/V	7.5 μA	C_{A5}	4 pF
G_7	200 μA/V	55 μA	C_{A3} & C_{A4}	4 pF
Bias		4 μA		
Osc		10 μA		

Current consumption

The amplifier as a whole is designed to consume 300 μA of current. In table 5-3 the value of the transconductance of each stage and the current consumption of the stage as well as the values of the capacitors are given. The bias current and oscillator consume 4 and 10 μA, respectively.

Fig. 5-22 *Chip micrograph.*

5.3.3 Measurement results

The design was implemented in a 0.7 μm CMOS process, and the chip micrograph is shown in figure 5-22. The chip area is 2.7 mm^2. The measured supply current is 280 μA from a 5.5 V supply voltage.

Noise Spectrum

The noise spectrum was measured at the output of the amplifier. For these measurements, the amplifier was used in an inverting configuration with a DC gain of 40 dB. The spectrum is shown in figure 5-23. The measurement shows that the input referred noise is 33 nV/√Hz. The $1/f$ noise is effectively suppressed by the combination of the chopper and auto-zero offset stabilization technique.

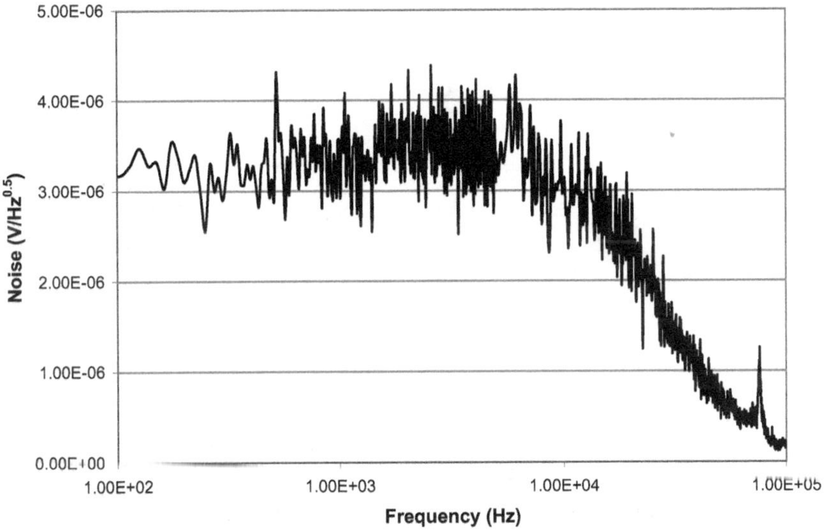

Fig. 5-23 *Spectrum of the output noise (gain 40 dB) showing that the noise is 3.3 μV/√Hz, indicating that the input referred noise is 33 nV/√Hz.*

Offset

In figure 5-24 the offset performance of two samples is shown versus the common-mode input voltage V_{CMin} at supply voltages V_{dd}=3.3 V and V_{dd}=5.5 V. From this measurement it can be concluded that the CMRR is 125 dB and the PSRR is 118 dB.

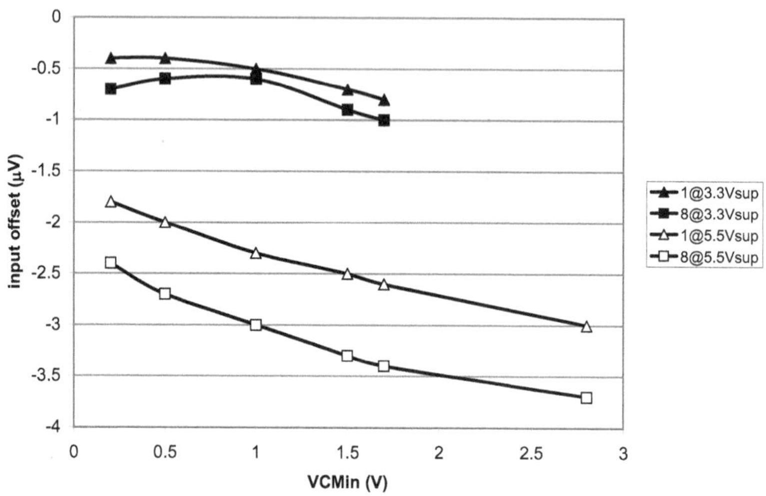

Fig. 5-24 *Input referred offset versus input common-mode voltage VCMin measured at two different supply voltages for two devices.*

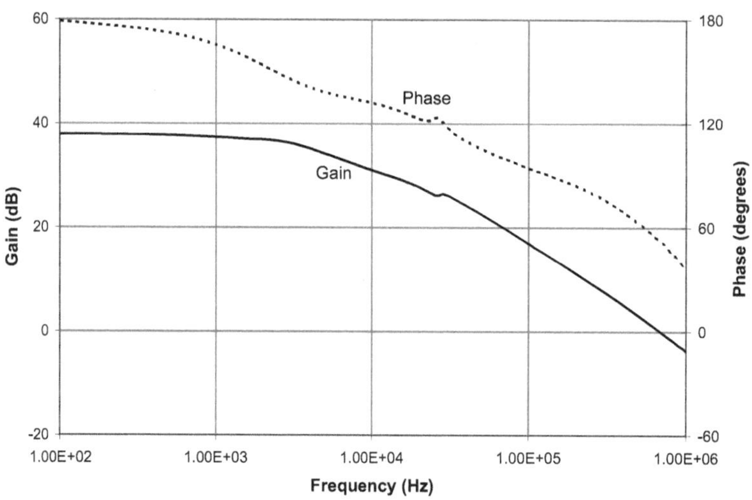

Fig. 5-25 *Bode plot of the operational amplifier in an inverting amplifier configuration with a DC gain of 40 dB.*

Frequency response

A measured Bode diagram is shown in figure 5-25. The measured unity gain frequency is 620 kHz with a phase margin of 57°. A chopper artifact can be noticed in the Bode diagram around 25 kHz. This artifact is caused by aliasing, caused by the sampling action of auto-zeroing.

5.4 Conclusions

In this chapter two operational amplifiers have been presented. The first is a chopper offset-stabilized operational amplifier, and the second is a chopper auto-zero offset-stabilized operational amplifier. Both circuits were implemented in a 0.7 μm CMOS process.

A chopper offset-stabilized operational amplifier with a GBW product of 1.3 MHz, a 140 nV/√Hz low frequency input referred noise and an offset of less than 1 μV at a 4 kHz chopper frequency, or an offset less than 1.5 μV at a chopper frequency of 16 kHz, was presented. The circuit consumes 700 μA from a 5.5V supply voltage.

This design proves that it is possible to design a low-offset wide bandwidth chopper offset-stabilized amplifier. It was shown that the use of a nested offset-stabilization loops significantly reduces the residual offset of the amplifier, and that the use of a sample-and-hold reduces the chopper ripple.

A chopper and auto-zero offset-stabilized amplifier was presented, which has a GBW product of 620 kHz, a 33 nV/√Hz low-frequency input referred noise, and an offset of less than 1.5 μV at a supply voltage of 3.3 V with a 25 kHz chopper frequency, and a 12.5kHz auto-zero frequency. The circuit consumes 280 μA from a 5.5 V supply voltage.

This design proves that a chopper auto-zero offset-stabilized operational amplifier is feasible. In comparison to the first design, in the second design the power consumption was reduced by 60%. Moreover, the noise level at low frequencies was also reduced, because of the use of a 50 times larger transconductance as stabilizing amplifier.

However, the lower noise also required a wider chopper switch to achieve a lower R_{on}. The resulting higher charge injection, together with the higher chopper frequency, leads to a higher residual offset.

5.5 References

[5.1] R. Burt, J.A. Zhang, "Micropower chopper-stabilized operational amplifier using a SC notch filter with synchronous integration inside the continuous-time signal path", *IEEE JSSC*, pp. 2729–2736, Dec. 2006.

[5.2] OPA333 Datasheet "1.8V, micropower CMOS operational amplifiers zero-drift series", www.ti.com, May 2007.

[5.3] J.H. Huijsing, M.J. Fonderie, "Chopper chopper-stabilized operational amplifiers and methods", US patent Nr. 6,734,723, Nov. 7, 2004.

[5.4] J.H. Huijsing, "*Operational amplifiers theory and design*", Dordrecht: Kluwer, 2001.

[5.5] D.M. Monticelli, "A quad CMOS single-supply opamp with rail-to-rail output swing", *IEEE JSSC*, Vol. SC-21, pp. 1026–1034, Dec. 1986.

[5.6] D. Neamen, "*Semiconductor Physics & Devices*", Second ed., New York, McGraw-Hill, 1997.

[5.7] K. Bult, G.J.G.M. Geelen, "A fast settling CMOS op amp for SC-circuits with 90-dBDC gain", *IEEE JSSC*, pp. 1379–1383, Dec. 1990.

Realizations of Instrumentation Amplifiers

6

6.1 Introduction

In chapters 2 and 3, the theory necessary for designing low-offset CMOS amplifiers was presented. In the previous chapter, implementations of operational amplifiers designed using this theory were discussed. This chapter focuses on realizations of the indirect current-feedback instrumentation amplifiers discussed in chapter 4.

This chapter focuses on two realizations of indirect current-feedback instrumentation amplifiers. The first is a chopper and auto-zero offset-stabilized indirect current-feedback instrumentation amplifier. The second is an indirect current-feedback instrumentation amplifier for high-side current-sense amplifier applications.

6.2 Low-offset indirect current-feedback instrumentation amplifier

6.2.1 Introduction

In this section of the book, a chopped and auto-zeroed offset-stabilized indirect current-feedback instrumentation amplifier will be presented. This design was implemented after the design which will be presented in section 6.3, and in parallel with the design presented in section 5.3. The purpose of this design is to investigate the performance of a chopper and auto-zero offset-stabilization topology for an instrumentation amplifier as discussed in chapter 4. This design is very similar to the design presented in section 5.3, except that extra input stages have been implemented to achieve an indirect current-feedback instrumentation amplifier.

First, the topology of the amplifier is presented. After that the transistor implementations of the used transconductances are discussed. Finally the measurement results are shown.

6.2.2 Topology

The simplified block diagram of the amplifier is shown in figure 6-1. There are two signal paths: a high-frequency path and a low-frequency path. The high-frequency path consists of input transconductor G_{21}, which amplifies the input voltage V_{in}, while feedback transconductor G_{22} amplifies the feedback-voltage V_{fb} across the resistor divider R_2 and R_1. This voltage is proportional to the output voltage V_{out}. The difference in their output currents drives the output stage G_1. The low frequency path consists of a chopped input transconductor G_{51}, which also amplifies V_{in}, and a chopped replica transconductor G_{52}, which also amplifies V_{fb}. The difference in their output currents drives an integrator around G_4, which in turn drives a transconductance G_3. If the transconductances of G_{51} and G_{52} are equal, the integrator's output will converge until V_{in} and V_{fb} are equal. When V_{in} and V_{fb} are equal, transconductance G_3 compensates for the offset of G_{21} and G_{22} by applying a current I_3.

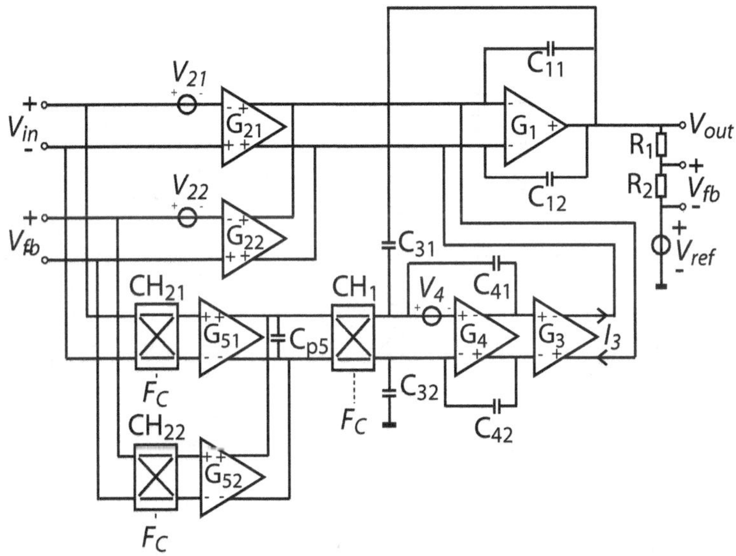

Fig. 6-1 *Simplified topology of a chopped offset-stabilized current-feedback instrumentation amplifier.*

Input stages G_{51} and G_{52} determine the low-frequency characteristics. The DC gain of the instrumentation amplifier can be expressed as:

$$\frac{V_{out}}{V_{in}} = \frac{R_1 + R_2}{R_2}\frac{G_{51}}{G_{52}} \approx \frac{R_1 + R_2}{R_2}, \tag{6-1}$$

assuming that G_{51} and G_{52} are identical.

Since there are two input stages in this implementation, the offsets of G_{51} and G_{52} create twice the chopper ripple compared to the operational amplifier implementations discussed in chapter 5. In order to reduce the chopper ripple in this design, a combination of both chopping and auto-zeroing was used, as discussed in section 5.3. This implementation, which is shown in figure 6-2, is similar to the operational amplifier shown in figure 5-15.

During the auto-zero phase F_{az}, both the inputs of G_{51} and G_{52} are shorted, and both transconductance stages are put into a unity gain configuration. During this phase the input offset voltages are stored on C_{A1} to C_{A4}.

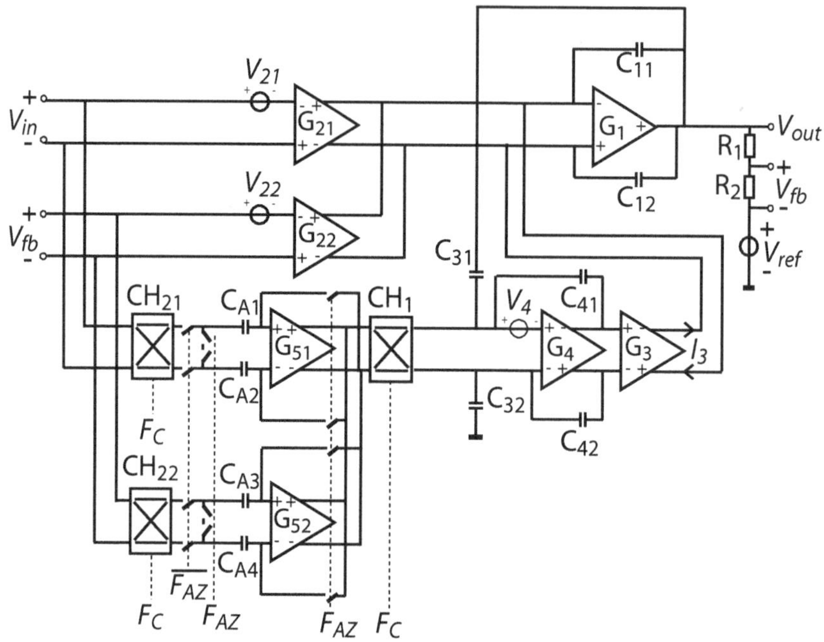

Fig. 6-2 *Simplified topology of a chopped auto-zeroed offset-stabilized current-feedback instrumentation amplifier.*

As discussed before in section 5.3, the offset of the integrator V_4 together with the parasitic capacitance C_{p5} seen at the inputs of CH_1, again, is a cause of residual offset. This residual offset is given by:

$$V_{ores} = \frac{4V_4F_CC_{p5}}{G_5} . \tag{6-2}$$

Therefore, an auto-zero offset-stabilization technique is used around the integrator, as shown in figure 6-3. The transconductance G_7 is auto-zeroed with clock $\overline{F}_1$. During clock F_1, the auto-zeroed transconductance G_7 measures the offset of integrator G_4 and applies current to capacitor C_{A5},

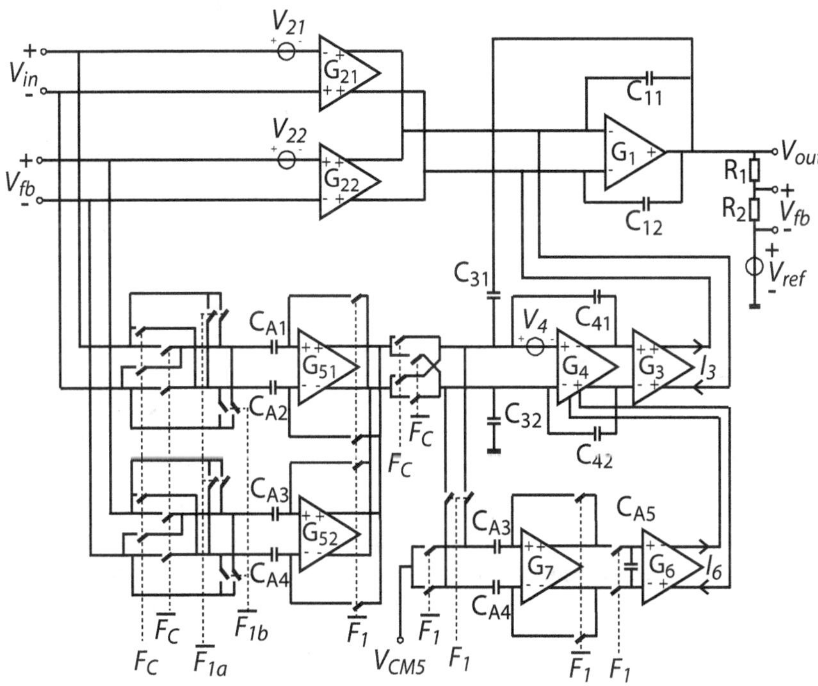

Fig. 6-3 *Actual implementation of the chopped auto-zeroed offset-stabilized instrumentation amplifier with auto-zero offset-stabilized integrator.*

which functions as a passive integrator. Stage G_6 is an auxiliary input of G_4. The offset V_4 of integrator transconductance G_4 can be reduced down to:

$$\frac{V_{4res}}{V_4} = \frac{G_4}{A_{o7}G_6}, \tag{6-3}$$

where A_{o7} is the DC voltage gain of G_7. Transconductors G_4 and G_6 share the same output impedance. The timing diagram is shown in figure 6-4.

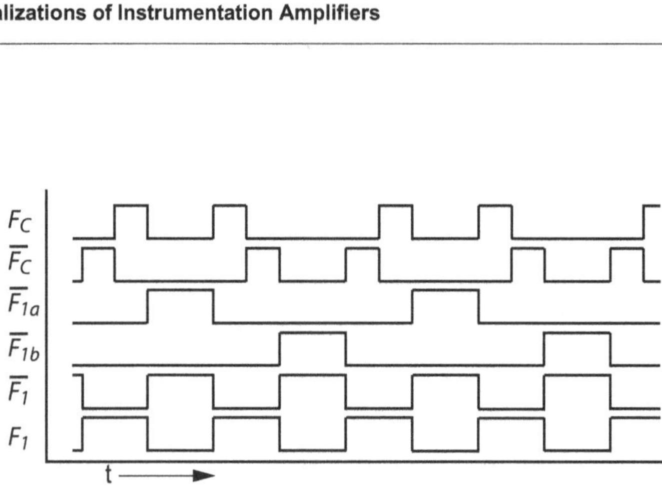

Fig. 6-4 *Timing diagram.*

6.2.3 Circuits

The circuits are very similar to the circuits shown in the operational amplifier implementation of chapter 5.3. The only difference is that there are additional input stages to implement the feedback and replica inputs of the indirect current-feedback amplifier. The implementations of G_1, G_{21}, G_{22}, and G_3 are shown in figure 6-5. The implementations of G_{51} and G_{52} are shown in figure 6-6.

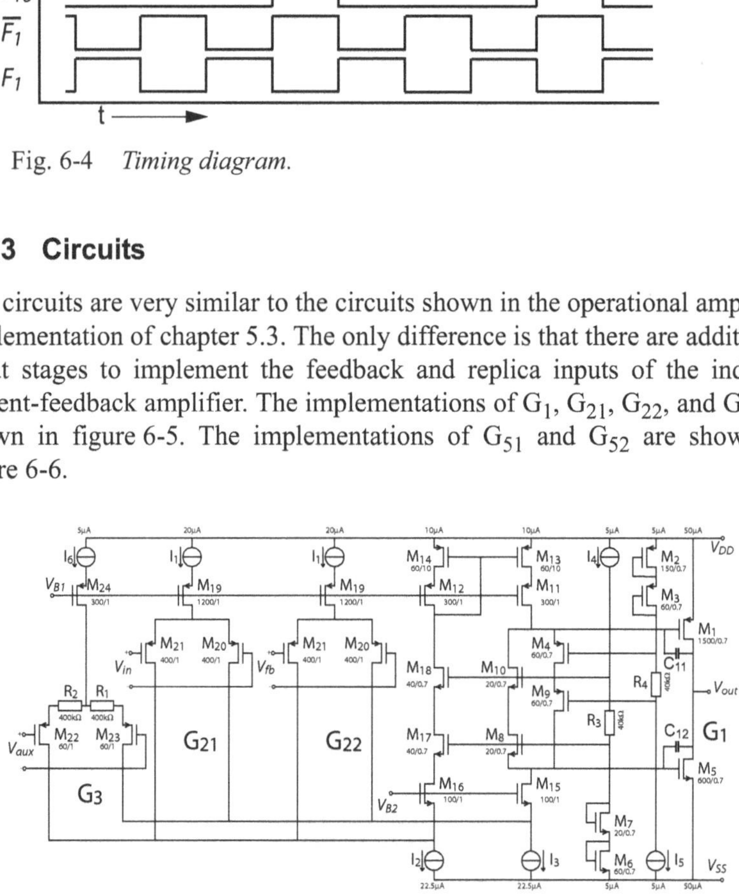

Fig. 6-5 *Implementation of G_1, G_{21}, G_{22}, and G_3.*

Table 6-1. *Transconductance and current consumption of each stage.*

Stage	Transcon-ductance	Current consumption	Capacitors	Cap value
G_1	1 mA/V	55 μA	$C_{11}+C_{12}$	32 pF
$G_{21}+G_{22}$	220 μA/V	75 μA		
G_3	2.5 μA/V	7.5 μA	C_{31} & C_{32}	32 pF
G_4	200 μA/V	50 μA	C_{41} & C_{42}	24 pF
$G_{51}+G_{52}$	220 μA/V	75 μA	C_{A1} & C_{A2}	4 pF
G_6	50 μA/V	7.5 μA	C_{A5}	4 pF
G_7	200 μA/V	55 μA	C_{A3} & C_{A4}	4 pF
Bias		4 μA		
Osc		10 μA		

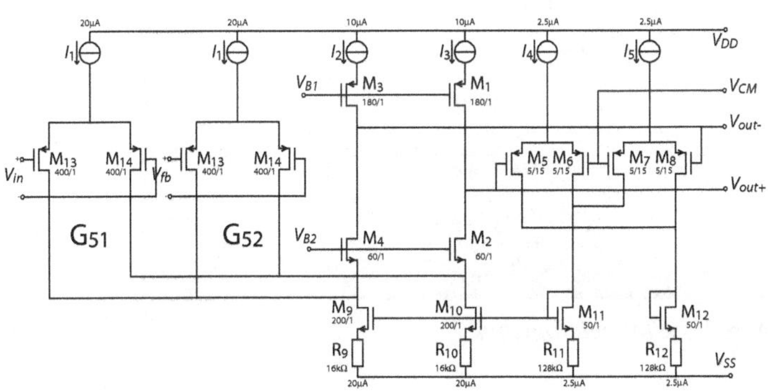

Fig. 6-6 *Implementation of G_{51} and G_{52}.*

The integrator transconductor G_4 is implemented as shown in figure 5-20, while the transconductor G_7 is implemented is shown in figure 5-19.

Current consumption

The amplifier as a whole is designed to consume 340 μA of current. In table 6-1, for each stage both the value of the transconductance and the current consumption of the stage as well as the values of the capacitors are given.

6.2.4 Measurement results

The design was implemented in a 0.7 μm CMOS process, and the chip micrograph is shown in figure 6-7. The chip area is 3.5 mm^2. It could have been 2.8 mm^2, if the layout of the internal transconductance stages had not been optimised for the operational amplifier design presented in figure 5.3. The measured supply current is 325 μA from a 5.5 V supply voltage.

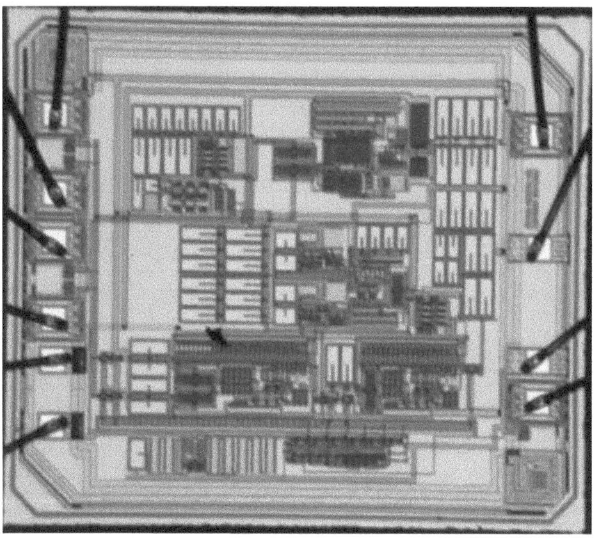

Fig. 6-7 *Chip micrograph.*

Noise

The noise spectrum was measured at the output of the amplifier. For these measurements, the amplifier was used in an inverting configuration with a DC gain of 40 dB. The spectrum is shown in figure 6-8. The measurement shows that the input referred noise is 42 nV/√Hz. Compared to the operational amplifier described in section 5.3 the noise increases with a factor 1.3. The noise seems flat and the $1/f$ corner is below 10 Hz.

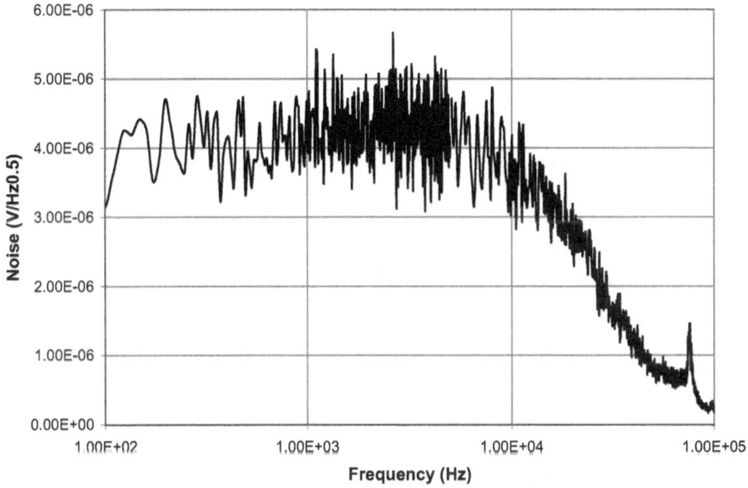

Fig. 6-8 *Spectrum of the output noise (gain 40 dB) showing that the noise is 4.2 $\mu V/\sqrt{Hz}$, which indicates that the input referred noise is 42 $nV/\sqrt{Hz}$.*

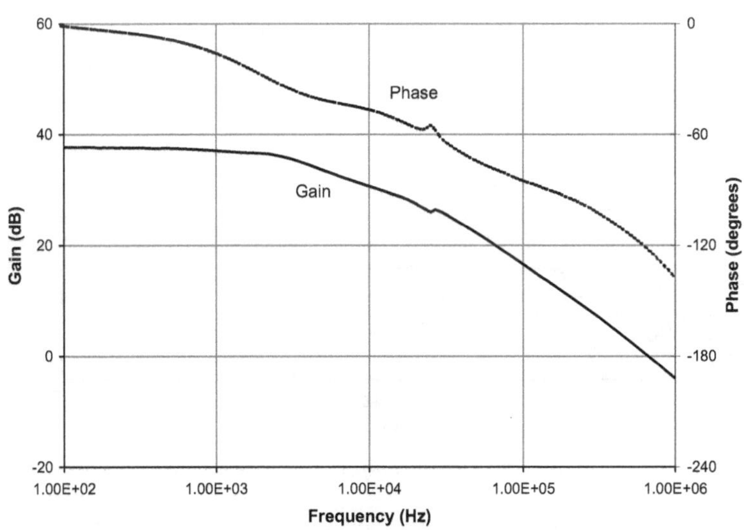

Fig. 6-9 *Bode plot of the operational amplifier in an inverting amplifier configuration with a DC gain of 40 dB.*

Frequency response

A measured bode plot is shown in figure 6-9. The measured unity gain frequency is 630 kHz with a phase margin of 57°. Around 25 kHz a chopper artifact can be observed. This artifact is caused by aliasing, caused by the sampling action of auto-zeroing.

Offset

The offset performance as a function of the supply voltage is shown in figure 6-10. From this figure it can be seen that the PSRR is 114 dB. In figure 6-11 the offset performance is shown versus the common-mode reference voltage V_{ref} at supply voltage V_{dd}=3.3 V and V_{dd}=5.5 V. In figure 6-12 the offset performance is shown versus the common-mode input voltage V_{CMin} at supply voltage V_{dd}=3.3 V and V_{dd}=5.5 V. It can be seen that the offset voltage is less than 2.5 µV for a supply voltage of 3.3 V. The CMRR for both V_{ref} and V_{CMin} is 130 dB for V_{dd}=5.5 V.

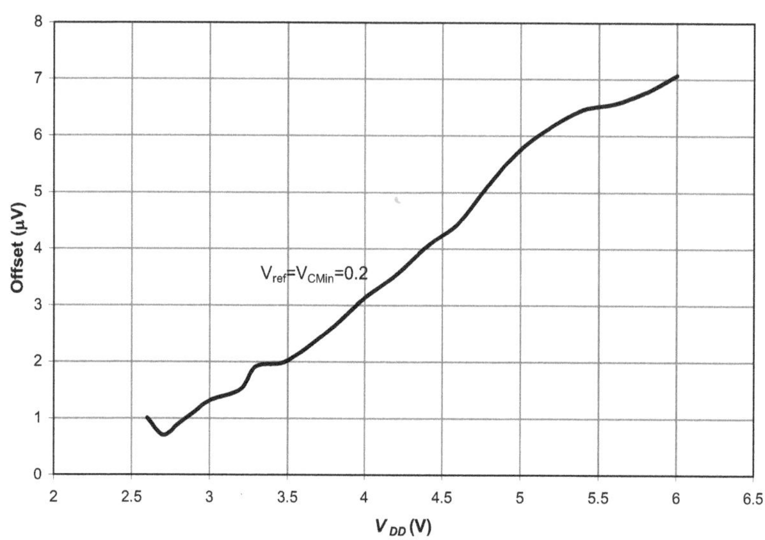

Fig. 6-10 *Offset as a function of supply voltage* V_{DD}.

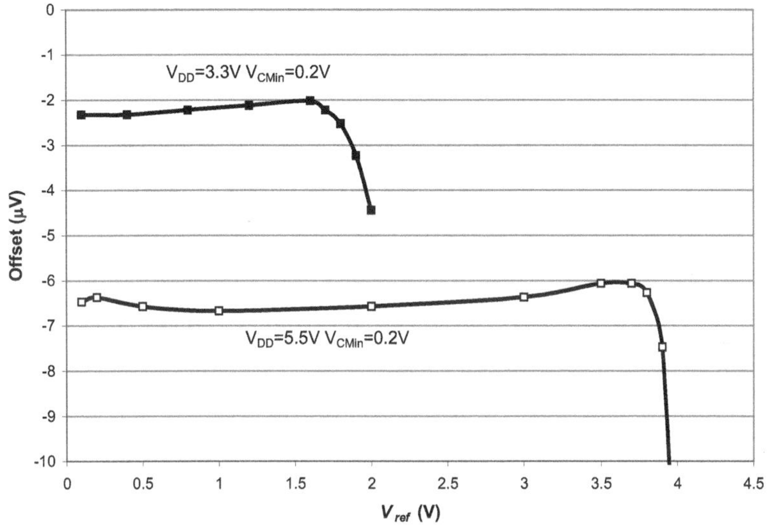

Fig. 6-11 *Offset as a function of reference voltage V_{ref} at two different supply voltages.*

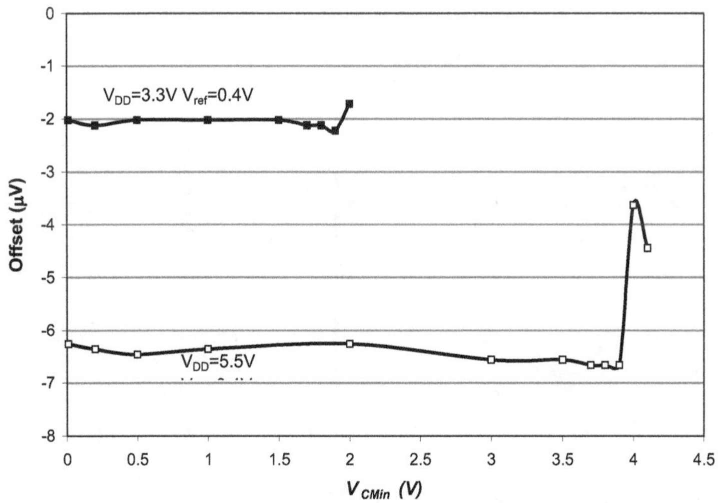

Fig. 6-12 *Offset as a function of the input common-mode voltage VCMin at two different supply voltages.*

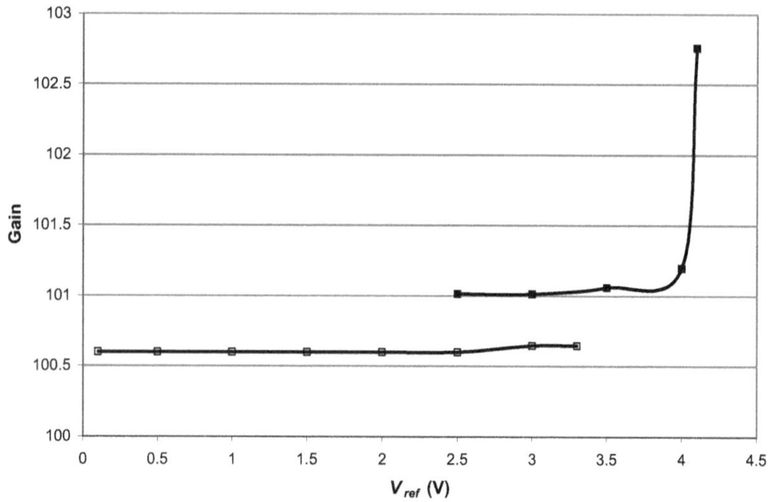

Fig. 6-13 *Gain as a function of the reference voltage* V_{ref}

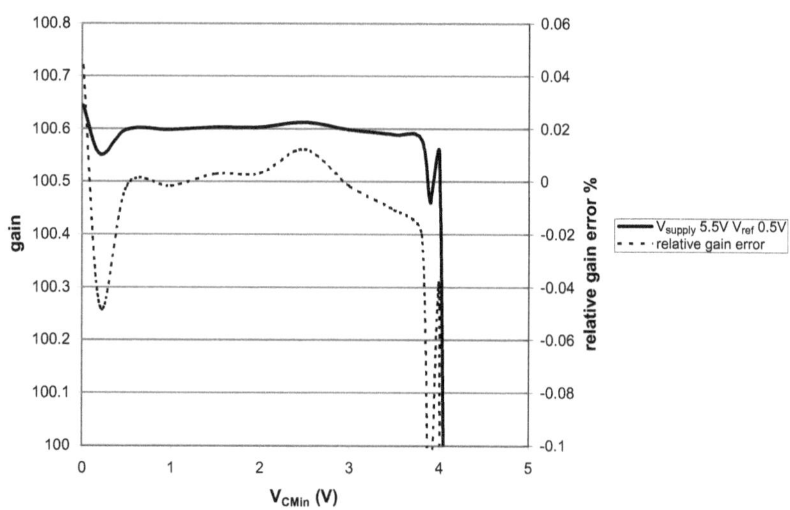

Fig. 6-14 *Gain as a function of the input common mode voltage* V_{CMin}.

Gain accuracy

In a current-feedback instrumentation amplifier, the value of the input transconductance can change over the input common-mode voltage, causing gain errors. Therefore, not only are offset, CMRR and PSRR important, but also the gain accuracy is important over the entire input common-mode range and reference common-mode range.

In figure 6-13 the gain is shown as a function of reference common-mode voltage V_{ref}. A discrepancy in gain can be seen, because the measurement setup has offset, and the input signal had to be inverted to prevent the output from clipping. In figure 6-14 the gain is shown as a function of input common-mode voltage V_{CMin}. It can be seen that the relative gain error is smaller than 0.05% from 0 to 3.6 V.

6.3 High-side current-sense amplifier

In this part of the book an indirect current-feedback instrumentation amplifier will be discussed for high-side current-sense applications [6.1]. First, current-sensing will be introduced. Afterwards, the topology of the used amplifier will be discussed in section 6.3.2. The circuits used in the implementation of this topology will be shown in section 6.3.3. Measurement results will be presented in section 6.3.4, followed by conclusions.

6.3.1 Current-sensing

Sensing supply currents is a fundamental requirement in many electronic systems, and the techniques to do so are as diverse as the applications themselves. Typical applications include: over-current protection, programmable current sources, and Coulomb counting to monitor the charge level of a battery.

The supply current is typically measured or sensed through a small current-sense resistor in series with the battery and load. So basically there are two options to implement this current-sense resistor R_S. It can either be implemented between the negative power supply and the load, which is called low-side current-sensing, or at the positive power supply, which is called the high-side current-sensing, both of which are depicted in figure 6-15. The voltage across the sense resistor will be amplified by a current-sense

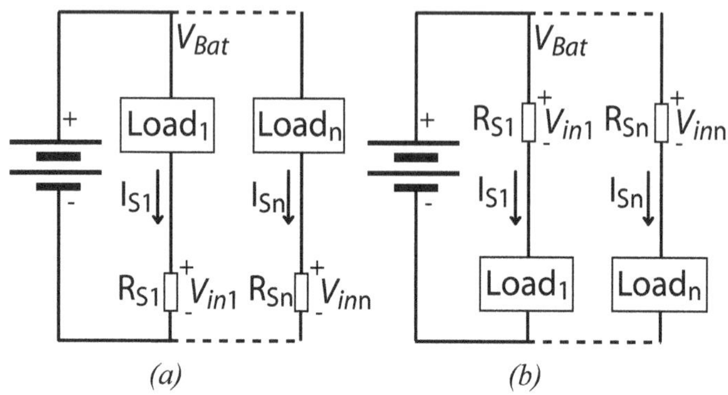

Fig. 6-15 *Current-sensing principle (a) low-side, (b) high-side.*

amplifier. In many applications it is also preferable to measure the supply currents through multiple loads.

Because of the low input common-mode voltage of the low-side current-sense amplifier, a single supply instrumentation amplifier could be used to amplify this signal. However, a high-side current-sense amplifier should also be able to withstand a high input common-mode voltage and refer its output voltage to ground. The implementation presented in this section focuses on the challenges of designing a high-side current-sense amplifier.

However, there are important system-level disadvantages of low-side current-sensing [6.2]. Firstly, the load is shifted from ground, which could be a problem in a system design with multiple parallel current-sense systems. Secondly, various fault mechanisms cause a load to be shorted to ground, which bypasses the current-sense resistor and cause currents to remain undetected by low-side current-sensing. These facts increase the demand for high-side current-sense amplifiers. A few specifications can be derived by focusing on the application of measuring the current flowing through a laptop-battery. Nowadays battery voltages already range up to 15 V, but they are expected to increase in the future. Moreover, a higher voltage is applied over the battery while it charges. Therefore it is not outrageous to design for a 30 V maximum input common-mode voltage. After all, the current flowing through the laptop battery can range from a stand-by current of approximately 10 mA to high-power currents of 10 A.

Generally speaking, to minimize the sense-resistor's value and its power dissipation, two specifications for high-side current-sense amplifiers

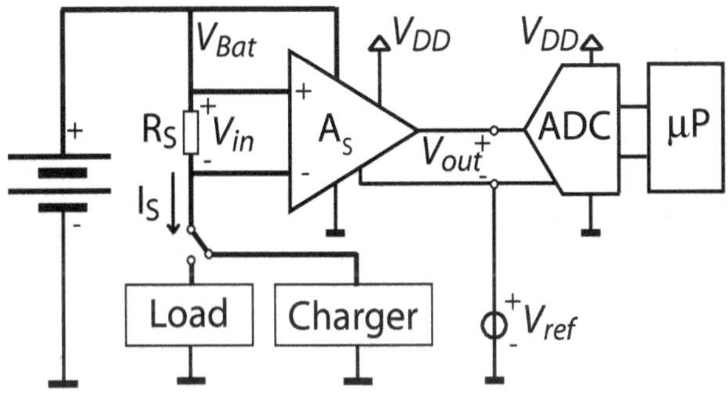

Fig. 6-16 *Bidirectional high-side current-sense system.*

are critical: input offset voltage and common-mode rejection ratio (CMRR). In a laptop-battery application, a sense resistor of 1 mΩ means that the input voltage V_{in} can range from 10 μV to 10 mV. Therefore, the input offset should be smaller than 10 μV and the CMRR should be higher than 130 dB to keep the offset below 10 μV over a 30 V range.

The circuit shown in figure 6-16 can be used to monitor the charge level of a battery. The current-sense amplifier A_S monitors the bidirectional current through the battery. This amounts to $V_{out}>V_{ref}$ for load currents and $V_{out}<V_{ref}$ for charge currents. A user can select V_{ref} to optimise the output dynamic range of the current-sense amplifier for its application; if there is no interest in the charging currents V_{ref} can be equal to ground. The ADC digitizes the output of the current-sense amplifier A_S. A micro-processor then integrates this value and stores a value, which is proportional to the charge in the battery. This application requires the current-sense amplifier to have a gain error of less than 0.5%. Moreover, because the laptop can consume power in spurs, a signal bandwidth of 1 kHz is also required.

Many topologies can be used for implementing a current-sense amplifier: a resistor bridge amplifier, a flying or switched capacitor amplifier, a current follower, or an instrumentation amplifier implementation. The last two are shown in figure 6-17.

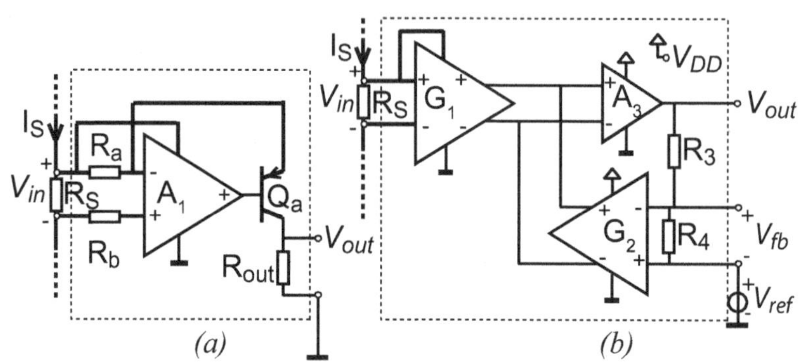

Fig. 6-17 *High-side current-sense amplifier based on (a) a V-to-I converter, (b) an indirect current-feedback instrumentation amplifier.*

In figure 6-17a, a current follower topology is used [6.3]. The gain of operational amplifier A_1 forces the sense voltage V_{in} onto the input resistor R_a. The current through this resistor equals the output current. For positive V_{in}, the gain of the current-sense amplifier can be expressed as:

$$\frac{V_{out}-V_{ref}}{V_{in}} = \frac{R_{out}}{R_a}\,. \tag{6-4}$$

Transistor Q_a isolates the high-voltage input from the low-voltage output, but offers no output voltage limit. To achieve a low-offset current-sense amplifier, the operational amplifier A_1 can be designed for low offset, as discussed in the previous chapters.

A topology based on an indirect current-feedback instrumentation amplifier is shown in figure 6-17b. By making use of current summing at the outputs of G_1 and G_2, the input and output, or reference common-mode voltage can be isolated [6.4][6.5]. This implies that the input common-mode voltage can be even lower than the output, or reference common-mode voltage. The input transconductance G_1 amplifies the input voltage V_{in}, while a replica transconductance G_2 amplifies the feedback-voltage V_{fb} across resistor divider R_3 and R_4. The difference in their output currents drives an operational amplifier A_3. The feedback cancels the output currents of G_1 and G_2. The gain of the current-sense amplifier is:

$$\frac{V_{out}-V_{ref}}{V_{in}}=\frac{R_3+R_4}{R_4}\frac{G_1}{G_2}. \tag{6-5}$$

The offset of the amplifier shown in figure 6-17b is the sum of the offsets of both G_1 and G_2. The next part of the book presents a low-offset indirect current-feedback instrumentation amplifier for high-side current-sensing applications. V_{DD} can range from 2.8 to 5.5 V, while the input common-mode V_{Bat} can independently range from 2 to 30 V by making use of current summing at the outputs of G_1 and G_2.

The use of separate supply voltages simplifies the task of interfacing the current-sense amplifier with other systems, for instance an ADC. Furthermore, the amplifier's output can be referred to an external reference voltage V_{ref}, which can range from 0 V to V_{DD}–1.4 V. Chopping and auto-zeroing are used to achieve an offset voltage of less than 5 μV over a CM input voltage range of 28 V, which in turn achieves a CMRR of more than 140 dB. Compared to a recently announced low-offset current-sense amplifier [6.6], the chosen topology enables bidirectional current-sensing, e.g. of the current through a rechargeable battery. Furthermore, trimmed gain-setting resistors are used to achieve 0.1% gain accuracy.

6.3.2 Topology

The simplified block diagram of the amplifier is shown in figure 6-18. A three-stage amplifier serves as the basis of this implementation. There are two signal paths. The low frequency path consists of a chopped input transconductor G_7, which amplifies the input voltage V_{in}, while a chopped feedback transconductor G_8 amplifies the feedback-voltage V_{fb} across the resistor divider R_{fb} and R_1. This voltage is proportional to the output voltage V_{out}.

The difference in the output currents of G_7 and G_8 drives an integrator around G_6, which in turn drives a transconductance G_5. If the transconductances of G_7 and G_8 are equal, the integrator's output will change until V_{in} and V_{fb} are equal, i.e. until the main amplifier's offset is compensated for. Transconductor G_5 compensates for the offset of G_3 and G_4 by applying a current I_5.

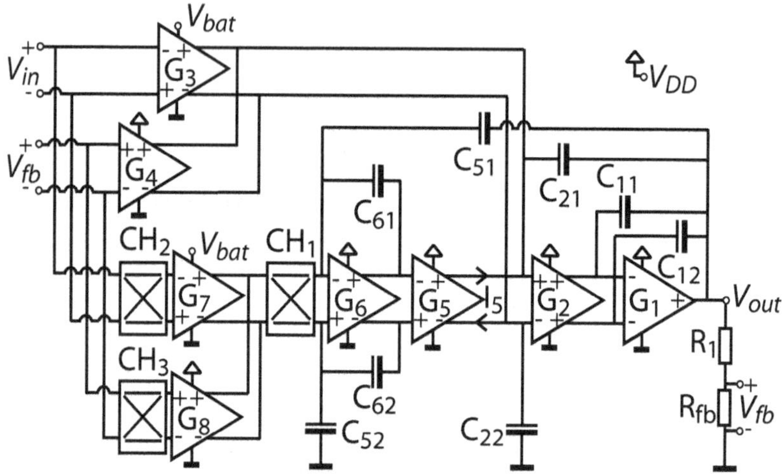

Fig. 6-18 *High-side current-sense amplifier based on a multi-path indirect current-feedback instrumentation amplifier.*

The high-frequency path consists of input transconductor G_3 and feedback transconductor G_4. The difference in their output currents drives a two-stage class-AB operational amplifier (G_2 and G_1). G_3 and G_7 are biased via the high-side CM input voltage V_{Bat}, while the other stages are biased via the supply voltage V_{DD}.

Transconductors G_7 and G_8 determine the low-frequency characteristics such as offset, low-frequency or $1/f$ noise, DC CMRR, and DC gain error. When their transconductances are equal, the output currents of G_7 and G_8 will cancel due to the feedback and so:

$$\frac{V_{out}}{V_{in}} = \frac{R_1 + R_{fb}}{R_{fb}} \frac{G_7}{G_8} \approx \frac{R_1 + R_{fb}}{R_{fb}}. \qquad (6\text{-}6)$$

The offset and low-frequency noise is chopper modulated by chopper CH_1, while the input and feedback signals are chopper-modulated and demodulated by choppers CH_2, CH_3, and CH_1, respectively. This will lead to the associated offset and low-frequency noise reduction. The transconductors G_3 and G_4 determine the high-frequency characteristics such as unity gain frequency and high-frequency CMRR.

The modulated offset voltage of G_7 and G_8 gives rise to ripple in the form of a triangular wave at the output of G_6, which in turn, gives rise to a triangular wave at the output of the whole amplifier. The input referred peak-to-peak voltage of this triangular wave is given by:

$$V_{in\text{-}pp} = (V_7 + V_8)\frac{G_{7,8}G_5}{2F_{ch}C_6G_{3,4}}. \tag{6-7}$$

To reduce this ripple, the combination of auto-zeroing and chopping is used, as shown in figure 6-19, where the low-frequency path is shown in more detail.

All switches between V_{fb} and G_8 are implemented with NMOS transistors. All switches between V_{in} and G_8 are implemented with PMOS switches in a high-voltage epi-pocket. Controlled by the clock F_1, the offsets of G_7 and G_8 are auto-zeroed by shorting their inputs, connecting G_8 in a unity-gain configuration, and then storing the resulting voltage on capacitors C_{A1} and C_{A2}. In addition, G_7 and G_8 are chopped by clock F_C to modulate both the residual offset due to finite voltage gain and the under-sampled noise associated with auto-zeroing [6.7]. A level-shift circuit drives the high-side chopper switches connected between the input voltage V_{in} and G_7. The system timing diagram is shown in figure 6-20. The base frequency of the chopper clock is 30 kHz clock. A 5 bit pseudo-random algorithm is used to vary this frequency between 28 kHz and 32 kHz to prevent aliasing.

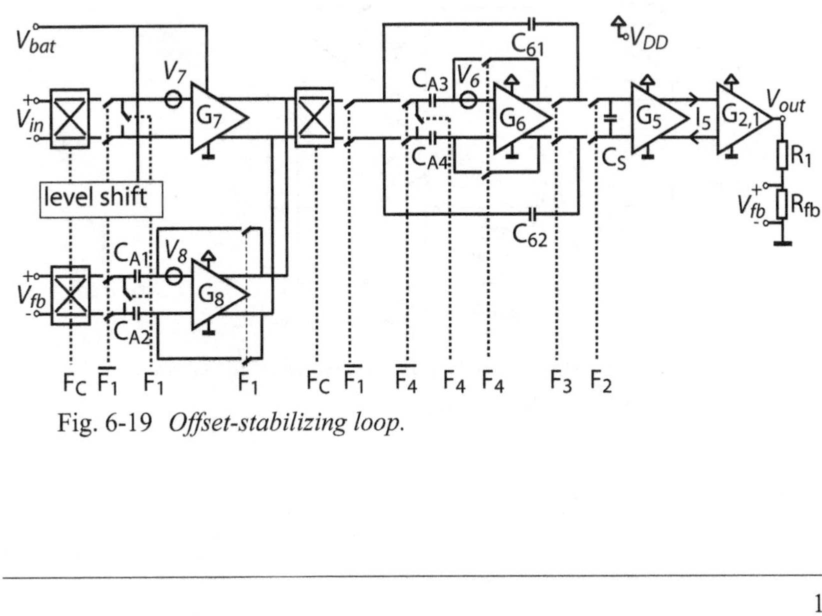

Fig. 6-19 *Offset-stabilizing loop.*

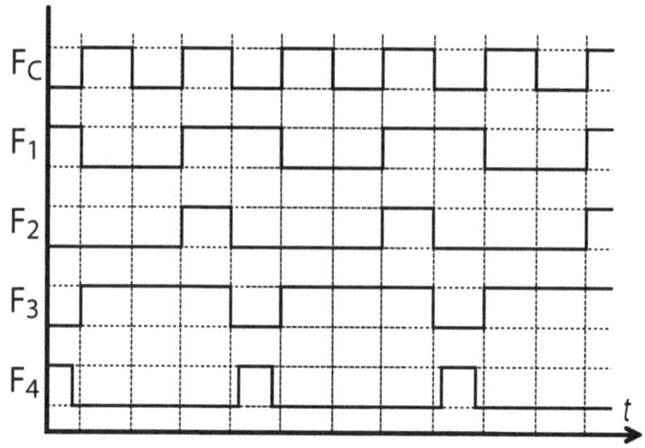

Fig. 6-20 *Timing diagram.*

Note that in this design there are no auto-zero capacitors connected between G_7 and the input voltage, as there are in the implementation of figure 6-3. This is because in the process used, no high-voltage capacitors were available. This, however, means that there is a systematic gain error in this implementation due to the parasitic capacitance at the inputs of G_8. This is illustrated in figure 6-21.

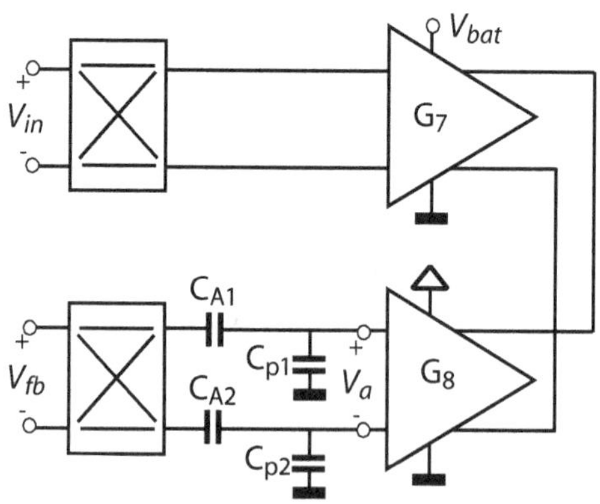

Fig. 6-21 *Input stages with gain error due to capacitive voltage division.*

The parasitic capacitances C_{p1} and C_{p2} model the parasitic capacitance caused by line capacitances. This causes the systematic voltage division:

$$\frac{V_a}{V_{fb}} = \frac{C_A}{C_p + C_A}, \tag{6-8}$$

where $C_A = C_{A1} + C_{A2}$ and $C_p = C_{p1} + C_{p2}$. This means that the overall gain of the instrumentation amplifier is:

$$\frac{V_{out}}{V_{in}} = \frac{R_1 + R_{fb}}{R_{fb}} \frac{G_7}{G_8} \frac{C_p + C_A}{C_A}, \tag{6-9}$$

which means that in order to maintain a gain which is accurately defined by the resistor values, G_7 should be equal to G_8, and C_a should be much larger than C_p.

The combination of auto-zeroing, chopping, and the multi-path technique is powerful. Auto-zeroing reduces the offset, and therefore the chopper ripple, while chopping modulates the folded noise associated with auto-zeroing to higher frequencies, where they can cause little harm because the characteristics of the amplifier are dominated by the high-frequency path. However, since the transconductances only see the input and feedback voltages half of the time, the signal-to-noise ratio is worsened by at least a factor $\sqrt{2}$.

The offset V_6 of the integrator also causes residual input offset [6.8]. To avoid this, the integrator is also auto-zeroed, as shown in figure 6-19. During the auto-zeroing of G_7 and G_8, the integrator's output voltage is sampled on C_S by clock F_2. This sampling operation also reduces the triangular ripple caused by chopping [6.9][6.8]. Next, the integrating capacitors C_{61} and C_{62} are disconnected from the output of G_6 by clock F_3, after which G_6 is configured in unity-gain and its offset V_6 will be stored on capacitors C_{A3} and C_{A4} by clock F_4. To avoid momentarily shorting the integration capacitors, F_3 and F_4 are designed to be non-overlapping clocks.

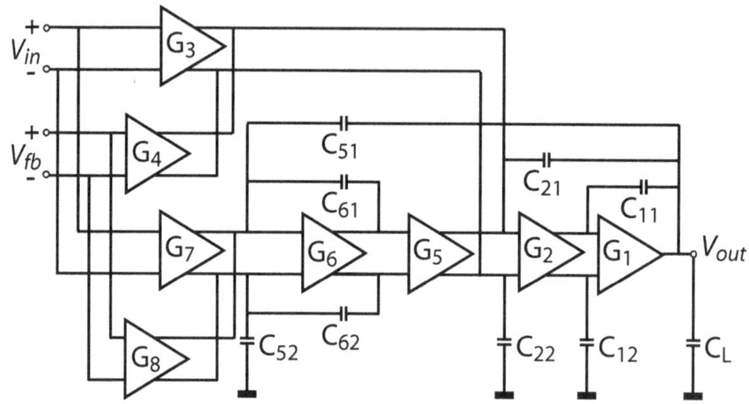

Fig. 6-22 *Multi-path hybrid-nested Miller frequency compensation in an indirect current-feedback instrumentation amplifier.*

In figure 6-22 the frequency compensation network is shown. Capacitors C_{11}, C_{12}, C_{21}, and C_{22} form a nested Miller compensation network [6.10], which is designed to obtain a GBW of 1MHz with a load capacitance C_L of 100pF. Capacitors C_{61} and C_{62} are used as integration capacitors. Capacitors C_{51} and C_{52} are used to implement a hybrid-nested Miller frequency compensation scheme with a smooth 20 dB/decade roll-off [6.10]. Without capacitors C_{51} and C_{52} the amplifier would only be conditionally stable.

6.3.3 Circuits

In this part of the book all the circuits are discussed that are used in this implementation. First, the gain determining input stages will be discussed. Secondly, the bias generator will be discussed, that generates the bias current for these input stages. Thirdly, the integrator implementation will be discussed. Lastly, the output stage will be discussed. In this design the W and L values of the transistors have been omitted because of third party interest.

Input stages

To sense the positive rail, the high-side input stages G_3 and G_7 should be designed with high-voltage capable NMOS input transistors. In contrast, the ground-sensing input stages G_4 and G_8 should be designed with PMOS input transistors. Since these stages use different types of transistors and are operated at different common-mode voltages, their transconductances will be inherently mismatched. To solve this problem, composite transistors are used, whose transconductances are set by resistors. In this way, accurate V-to-I converters can be realised [6.4] [6.11]. A simplified schematic of G_7 and G_8 is shown in figure 6-23.

The high-voltage input transconductors G_3 and G_7 are designed as follows. The input NMOS transistors M_1 and M_2 act as voltage followers and force the input voltage across resistors R_1 and R_2. Transistors M_1 and M_2 are implemented with regular low-voltage NMOS devices, which are biased with a high-voltage bulk. This can be done because a twin-well BiCMOS process is used.

The drains of M_1 and M_2 are connected to high-voltage PMOS folded cascodes M_3 and M_4, which drive high-voltage NMOS M_5 and M_6 functioning as inverting amplifiers and as current-followers. The high gain of this local loop ensures that the transconductance of G_3 and G_7 is accurately defined by the values of R_1 and R_2. With the biasing currents shown in figure 6-23, the amplifier's simulated gain variation is less than 0.15% over

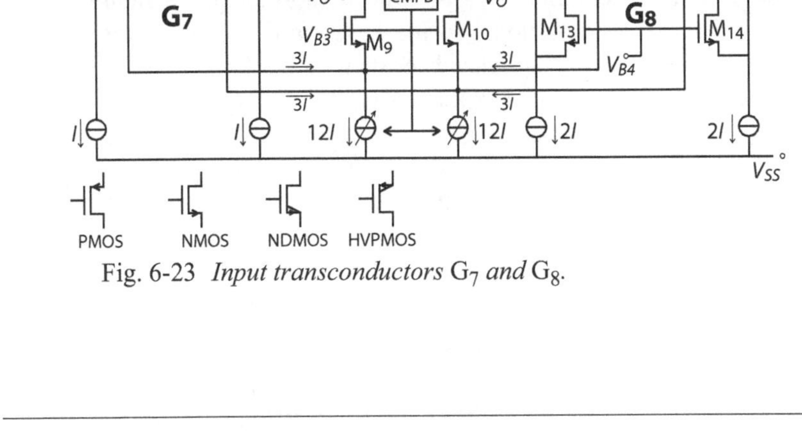

Fig. 6-23 *Input transconductors* G_7 *and* G_8.

the entire common-mode range. The high-voltage devices M_3 to M_6 can handle common-mode voltages as high as 30 V, which is the limit imposed by the process used. Transistors M_1 to M_4 are biased at a current *I*. As a result, the maximum input differential voltage range is *6IR*, where *R* is the value of R_1 or R_2. In this design this value corresponds to 150 mV.

A similar topology is used for G_4 and G_8, but with PMOS input transistors, NMOS cascodes, and NMOS current followers [6.11]. The input PMOS transistors M_{11} and M_{12} act as voltage followers and force the input voltage across resistors R_3 and R_4. The drains of M_{11} and M_{12} are connected to NMOS folded cascodes M_{13} and M_{14}, which drive NMOS transistors M_{15} and M_{16} functioning as inverting amplifiers and as current-followers. The high gain of this local loop ensures that the transconductance of G_4 and G_8 is accurately defined by the values of R_3 and R_4.

To define the gain of the indirect current-feedback amplifier by external resistors, the transconductance defining both resistors R_1 and R_2 in G_7 and R_3 and R_4 in G_8 are laser-trimmed.

The output currents of the input stages are summed together in a folded cascode. The fully differential amplifier needs one common-mode feedback loop. The transconductance is determined by the degeneration resistor. As previously mentioned, the maximum input differential voltage range is *6IR*. To keep this value constant over temperature, a current should be used, that has the same temperature characteristic as the degeneration resistors.

Bias generator

Contrary to the constant transconductance biasing used in all other amplification stages discussed in section 5.2.2, the degenerated input stages need to be biased with a constant current over temperature to maintain a constant differential input voltage range. The circuit used is shown in figure 6-24. In this circuit both a proportional to absolute temperature [PTAT] [6.12] and a current proportional to the base emitter voltage are summed together to generate a bandgap related constant current.

The PTAT current can be expressed as:

$$I_{ptat} = \frac{kT}{qR_2} \ln \frac{A_2 A_3}{A_4 A_1}, \tag{6-10}$$

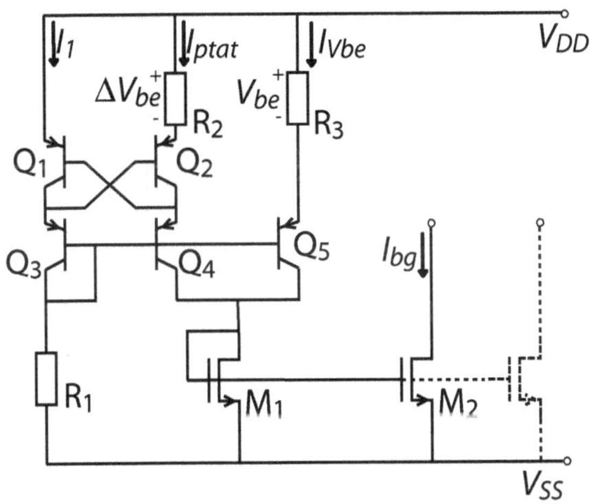

Fig. 6-24 *BiCMOS constant current generator.*

in which A_x is the emitter area of the indexed transistor. The V_{be} current can be expressed as:

$$I_{VBE}=\frac{V_{be5}-V_{be4}-V_{be1}}{R_3}\approx\frac{-V_{be1}}{R_3}=\frac{kT}{qR_3}\ln\frac{I_1}{I_{s1}}\ , \tag{6-11}$$

in which I_{s1} is the saturation current of Q_1. Since this parameter is highly temperature dependent, the current I_{VBE} has a negative temperature dependency. Furthermore, since the V_{be} and ΔV_{be} voltages have opposed temperature dependencies, an almost temperature-independent I_{bg} can be obtained by choosing the appropriate values for resistors R_2 and R_3.

Integrator

In figure 6-25 the implementation of the integrator transconductance G_6 is shown. An implementation very similar to figure 5-6 is used, the difference being that to assure a 100 dB open loop DC voltage gain, the cascode transistors M_9 and M_{10} are added to the input transistors M_3 and M_4. These cascode transistors are biased with diode connected transistor M_8. The common-mode feedback circuit is also implemented in the same way as in figure 5-6.

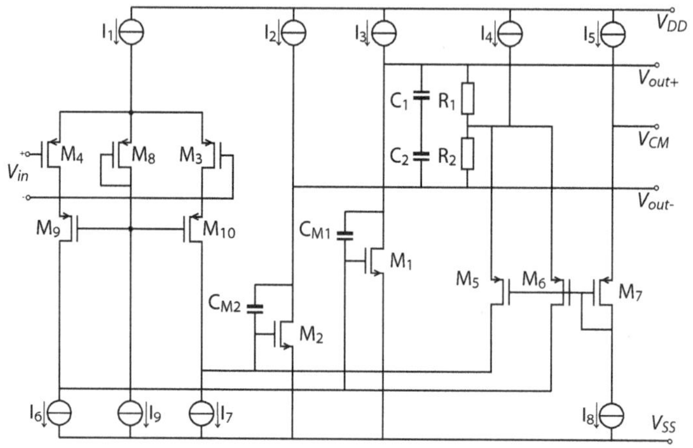

Fig. 6-25 *Two-stage class-A operational amplifier with input cascode.*

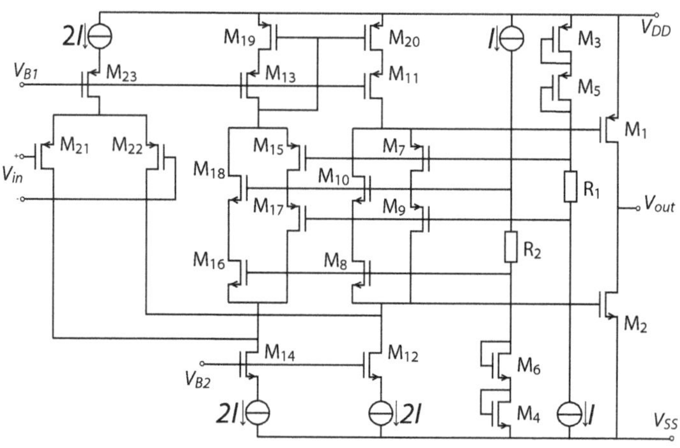

Fig. 6-26 *Two-stage class-AB operational amplifier with improved dependency of quiescent current to V_{DD}.*

A potential disadvantage of this circuit is the switching transient behaviour, caused by the two-stage topology. Since this amplifier is being auto-zeroed, the output jumps from the common-mode voltage to the value stored on the integrator capacitors. The two-stage topology has an increased settling time with respect to the single-stage topology. For this reason the folded cascode implementation was developed, which was later used in the already discussed implementations in sections 5.3 and 6.2.

Output stage

The output stages G_1 and G_2 in this topology are designed in the same way as the operational amplifier discussed in section 5.2, except that the class-AB bias is implemented again to obtain improved behaviour with respect to the power supply voltage. The circuit is shown in figure 6-26. It is the same circuit as discussed in section 5.3.2; figure 5-18, except that in order to keep the layout symmetrical, transistors M_{15} and M_{17} are added as dummy transistors.

6.3.4 Measurement results

The current-sense amplifier was fabricated in a 0.8 μm BICMOS process with high-voltage transistors and laser-trimmed thin-film resistors. It has a die area of 2.5 mm^2. The chip micrograph is shown in figure 6-27. The output noise (gain = 11) spectrum density is shown in figure 6-28. At frequencies below 10 kHz the input referred noise density is 136 nV/√Hz. At frequencies up to 15 kHz, a slight increase in noise can be seen which is due to both the combination of auto-zeroing and chopping, and the use of an offset-stabilizing topology [6.9][6.8][6.7]. After 15 kHz the noise goes down almost linearly with frequency to a value of 70 nV/√Hz.

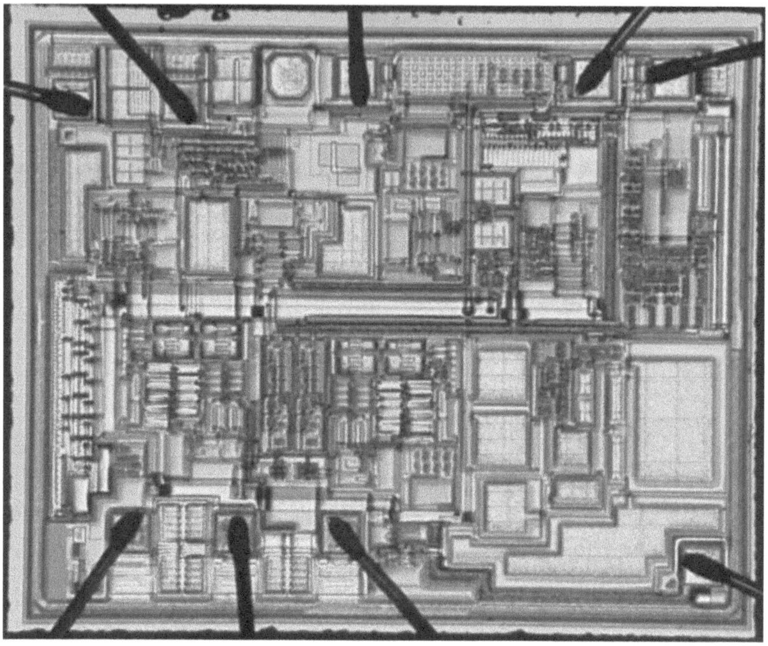

Fig. 6-27 *Chip micrograph.*

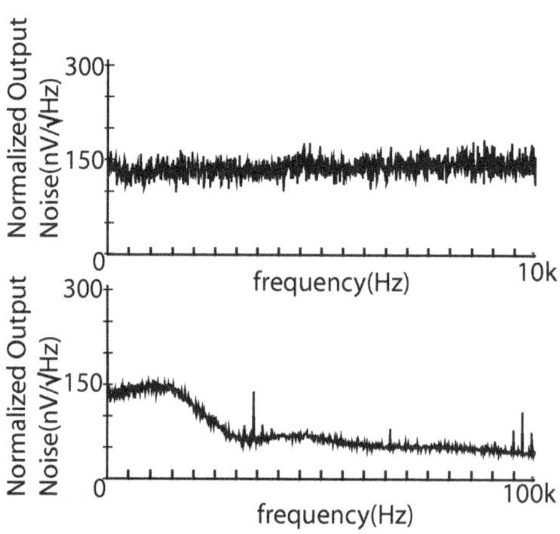

Fig. 6-28 *Output noise spectrum divided by the gain; gain = 11.*

Measurements of ten samples show that the amplifier's offset voltage is less than 5 μV, which together with its gain accuracy of 0.1% allows for precise current measurements even with small sense resistors. In figure 6-29 the offset performance of two samples is shown versus the reference voltage V_{ref} at V_{dd}=3.3 V, which is the common-mode voltage of the feedback input V_{fb}. It can be seen that the amplifier saturates when V_{ref}=2.2 V. The CMRR for the reference input is about 120 dB.

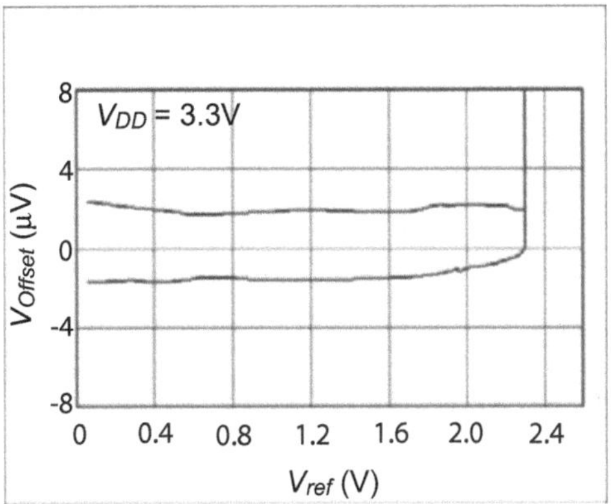

Fig. 6-29 *Input referred offset voltage versus* V_{ref}.

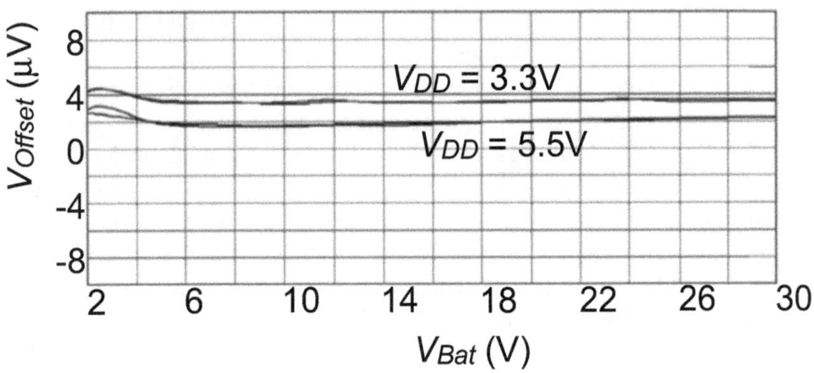

Fig. 6-30 *Input referred offset as a function of input common-mode voltage* V_{Bat}.

In figure 6-30 the offset performance of two samples is shown versus the input common-mode voltage V_{Bat}. It can be seen that the offset stays within 2 μV over a 28 V change in V_{Bat}, which refers to 143 dB CMRR. It can also be seen that the offset changes about 2 μV over a 2.2 V change in V_{DD}, which refers to a 121 dB PSRR. In figure 6-31 the CMRR is presented as a function of frequency. For frequencies below 20 kHz, the CMRR is more than 105 dB.

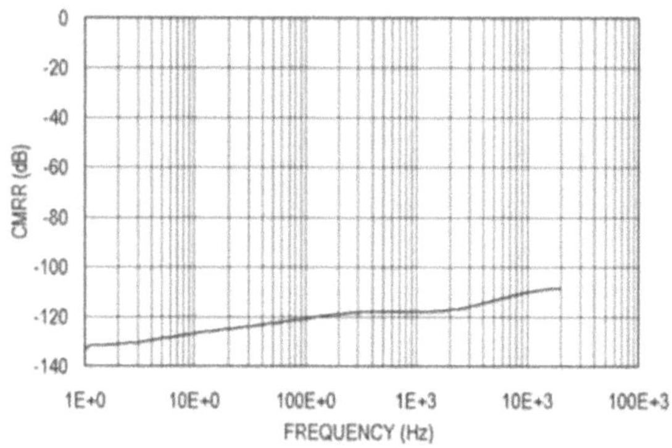

Fig. 6-31 *CMRR as a function of frequency.*

Not only offset is important for a current-sense amplifier, but, the gain accuracy is also important over both the entire input common-mode and reference common-mode range. In a current-feedback instrumentation amplifier, the value of the input transconductance can change over input common-mode voltage, causing gain errors.

In figure 6-32 the gain error is shown as a function of input common-mode voltage V_{Bat} at three reference voltages. Clearly, both voltages have an influence on the gain error. However, it can be seen that there is a ±0.15% gain error caused by increasing V_{Bat} from 2 to 28 V. A 0.1% gain error is caused by increasing V_{ref} from 0 to 1.9 V when V_{DD} is equal to 3.3 V.

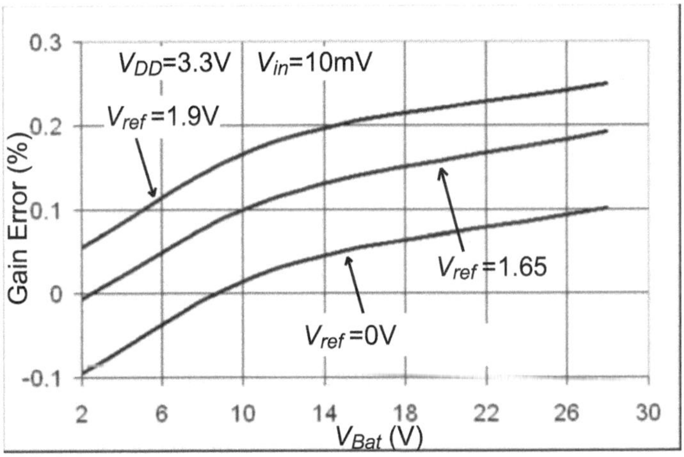

Fig. 6-32 *Gain error as a function of input common-mode voltage V_{Bat}.*

6.4 Conclusions

In this chapter two designs of chopper and auto-zero offset-stabilized indirect current-feedback instrumentation amplifiers were presented. The first was a 5.5V instrumentation amplifier, and the second was a current-sense amplifier, which is an instrumentation amplifier having one input fixed to a supply voltage which can range between 1.9 and 28 V.

Firstly, a chopper and auto-zero offset stabilized indirect current-feedback instrumentation amplifier was presented. The GBW is 630 kHz, and low-frequency input referred noise 42 nV/√Hz while consuming 325 μA from a 5.5 V supply voltage. The PSRR is 114 dB and the CMRR of both the reference and input common mode is 130 dB. The gain accuracy is within 0.5% over a reference voltage from 0 to 3.6 V at a 5.5 V supply voltage.

Secondly, a fully integrated indirect current-feedback instrumentation amplifier for current-sense amplifier applications was presented with an offset voltage of less than 5 μV. It has a common-mode input range from 2 V to 30 V and achieves a 143 dB DC CMRR. The reference has a common-mode range from 0 V to V_{DD}–1.4 V. This gives the current-sense amplifier the intrinsic bidirectional behaviour of an instrumentation amplifier. By using accurate V-to-I converters as input stages, a gain accuracy of 0.1% can be obtained for a fixed reference voltage. The GBW is 1 MHz, and a low-frequency input referred noise of 150 nV/√Hz is obtained, while consuming 1,050 μA supply current.

In table 6-2, this design is compared to a recently introduced current-sense amplifier [6.6]. These two amplifiers offer a new level of precision in current-sensing. In addition, however, the topology presented here has the natural bidirectional current-sensing capability of an instrumentation amplifier.

The first design has a better gain accuracy over common-mode voltage than the second design for two reasons. Firstly, the input stage and replica stage are identical in the first design, while in the second design they are completely different. Secondly, in the first design the input common-mode voltages seen by the gain setting stages are equal. Therefore, the gain accuracy is dependent on the quality of the capacitors and switches, in

Table 6-2. *Comparison of low-offset current-sense amplifiers.*

	LTC6102 [6.6]	This work
Year of release/ publication	2007	2008
Offset	10 μV	5 μV
Bidirectional	No	Yes
Gain Accuracy	External resistor dependent	0.15%
I_{supply} shutdown	No	1 μA
V_{DD}	V_{Bat}	2.8 to 5.5 V
V_{Bat}	4 to 60 V	1.9 to 30 V
V_{out}	0 to 8 V	0 to V_{DD}
I_{supply} from V_{Bat}	420 μA	200 μA
I_{supply} from V_{DD}		850 μA

contrast to the second design in which the high-voltage input operates at a different input common-mode voltage.

6.5 References

[6.1] J.F. Witte, J.H. Huijsing, K.A.A. Makinwa, "A current-feedback instrumentation amplifier with 5mV offset for bidirectional high-Side current-sensing", *IEEE ISSCC*, pp. 74–75, Feb. 2008.

[6.2] Linear Technology Corp., "Application note 105: Current sense circuit collection", http://www.linear.com/ad/current_sense.jsp, Feb. 2008.

[6.3] Linear Technology Corp., "LT1787", Datasheet, www.linear.com, Apr. 2007.

[6.4] R.J. v.d. Plassche, "A wide-band monolithic instrumentation amplifier", *IEEE JSSC*, pp. 424–431, Dec. 1975.

[6.5] B.J. van den Dool, J.H. Huijsing, "Indirect current feedback instrumentation amplifier with a common-mode input range that includes the negative rail", *IEEE JSSC*, pp. 743–749, July 1993.

[6.6] Linear Technology Corp., "LTC6102", Datasheet, www.linear.com, July 2007.

[6.7] A.T.K. Tang, "A 3 μV-offset operational amplifier with 20 nV/√Hz input noise PSD at DC employing both chopping and autozeroing", *IEEE ISSCC*, pp. 386–387, Feb. 2002.

[6.8] J.F. Witte, K.A.A. Makinwa, J.H. Huijsing, "A CMOS offset-stabilized opamp", *IEEE JSSC*, pp. 1529–1535, July 2007.

[6.9] R. Burt, J. Zhang, "A micropower chopper-stabilized operational amplifier using a SC notch filter with synchronous integration inside the continuous-time signal path", *IEEE JSSC*, pp. 2729–2736, Dec. 2006.

[6.10] J.H. Huijsing, "*Operational amplifiers theory and design*", Dordrecht: Kluwer, 2001.

[6.11] J.H. Huijsing, B. Shahi, "Accurate voltage-to-current converters for rail-sensing current-feedback instrumentation amplifiers", US patent Nr. 7,202,738, Apr. 10, 2007.

[6.12] R.C. Dobkin, "Input supply independent circuit", US patent Nr. 3,930,172, Dec. 30, 1975.
[6.13] Linear Technology Corp., "LT6104", Datasheet, www.linear.com, Apr. 2007.

Conclusions and Future Directions

7

7.1 Conclusions

The theory and realizations presented in this book show that the offset-stabilization techniques, as discussed in section 3.3, can be used to design broad-band low-offset general-purpose operational amplifiers and instrumentation amplifiers. The chopper offset-stabilization amplifier can be used to obtain good noise performance. However, the chopper ripple remains a problem. This problem can be overcome by using a sample-and-hold in order to low-pass filter the ripple [7.1] [7.2], or to use a chopper and auto-zero offset-stabilization amplifier [7.3]. The latter technique will lead to amplifiers with a $\sqrt{2}$ higher low-frequency noise, unless, a ping-pong technique is used [7.4] [7.5].

7.2 Future directions

The field of dynamic offset compensation techniques does not end with this book. A number of advancements can still be made in future research projects. This section gives just an idea of what the future will bring and what kinds of projects might follow the work presented in this book.

An important remark to be made is that there should always be a market for a product. For instance, an expensive amplifier with a less than 100 nV input-referred offset will probably not do very well, because there are simply not many applications that really benefit from an offset of less than 1 μV. Thermocouple effects on printed circuit boards would also ruin the ultra-low offset. These same thermocouple effects increase the testing cost of such amplifiers. Therefore, a general-purpose instrumentation amplifier with less than 1 μV offset will probably never be available on the market. However, if a sensor interface requires offsets of less than 100 nV, for instance in Hall sensors for compass applications [7.6], this requirement could be achieved by custom on-chip read-out electronics.

Recently a CMOS amplifier has been published in which trimming was used to obtain a low offset with a 0.33 μV/°C offset drift [7.7]. In chapter 2, it was shown that dynamic offset compensated amplifiers perform very well if the inherent offset is already low. In the future, there might be some products in which offset trimmed amplifiers are also chopped to obtain a sub-microvolt offset and low noise. If and when these circuits will be developed would depend on the market. If the increased trimming cost is believed to be more beneficial than the increased circuit complexity, which increases both design time and time to market, it will happen.

In this book the feasibility has been proven for a low-offset indirect current-feedback instrumentation amplifier for high-side current sensing applications. In the future, this topology could be adopted by the industry. Nowadays, high-voltage low-offset amplifiers are available for current sensing applications [7.3] [7.8]. In these applications, one input is always connected to the high-voltage supply. It might be an interesting project to design a high-voltage rail-to-rail input stage capable of sensing up to both the high-voltage rail as well as the ground rail. This way a current-sense amplifier could be obtained for both low-side and high-side current-sensing applications.

In this book two techniques has been discussed in which the chopper ripple in chopper amplifiers or chopper offset-stabilized amplifiers is reduced. The first technique uses a sampler which samples the chopper ripple. The second technique uses a chopped and auto-zeroed input stage which samples the input offset and thereby decreases the ripple. Both techniques, however, use a sampling action. This can cause a ripple due to kT/C noise of the sampling capacitors, noise aliasing, and signal aliasing effects. It is advisable to investigate future techniques in which the ripple can be reduced by using continuous-time circuits. A continuous time ripple reduction loop has already been described in [7.9].

The chopper offset-stabilized chopper amplifier and specifically the iterative offset-stabilization technique discussed in section 3.4.1 have the potential of a very low offset, since every source of offset has been offset-stabilized. This topology, however, can still be thoroughly researched. The ping-pong-pang instrumentation amplifier proposed in section 4.2.3 [7.10] has the potential to obtain a very good gain-error specification, which can also still be researched more thoroughly.

A step beyond dynamic offset compensation is dynamic offset and gain compensation. It might be advantageous to design an instrumentation amplifier that dynamically compensates for its own offset as well as its own gain error. It could, for instance, be possible to auto-calibrate two transconductances by applying the same input signal to both their inputs and auto-calibrating their output currents to be equal. Combining dynamic gain compensation with dynamic offset compensation could be another interesting research challenge.

7.3 References

[7.1] R. Burt, J. Zhang, "A micropower chopper-stabilized operational amplifier using a SC notch filter with synchronous integration inside the continuous-time signal path", *IEEE JSSC*, pp. 2729–2736, Dec. 2006.

[7.2] J.F. Witte, K.A.A. Makinwa, J.H. Huijsing, "A CMOS chopper offset-stabilized opamp", *IEEE JSSC*, pp. 1529–1535, July 2007.

[7.3] J.F. Witte, J.H. Huijsing, K.A.A. Makinwa, "A current-feedback instrumentation amplifier with 5μV offset for Bidirectional High-Side Current-Sensing", *IEEE ISSCC*, pp. 74–75, Feb. 2008.

[7.4] Texas Instruments, "0.05μV/C max, single-supply CMOS op amps zero-drift", Datasheet, Rev. D, OPA335, June 12, 2003.

[7.5] Thomas Kugelstadt, "Auto-zero amplifiers ease the design of high-precision circuits", Texas Instruments, 2005, focus.ti.com/lit/an/slyt204/slyt204.pdf.

[7.6] J.C. van der Meer, F.R. Riedijk, E. van Kampen, K.A.A. Makinwa, J.H. Huijsing, "A fully integrated CMOS Hall sensor with a 3.65μT 3σ offset for compass applications", *IEEE ISSCC*, pp. 246–247, Feb. 2005.

[7.7] M. Bolatkale, M.A.P. Pertijs, W.J. Kindt, J.H. Huijsing, K.A.A. Makinwa, "A biCMOS operational amplifier achieving 0.33μV/°C offset drift using room-temperature trimming", *IEEE ISSCC*, pp. 76–77, Feb. 2008.

[7.8] Linear Technology Corp., "LTC6102 data sheet", www.linear.com, July 2007.

[7.9] R. Wu, K.A.A. Makinwa, J.H. Huijsing, "A chopper current-feedback instrumentation amplifier with a 1 mHz 1/f noise corner and an AC-coupled ripple-reduction loop", IEEE ISSCC, pp. 322–323, Feb. 2009.

[7.10] J.H. Huijsing, "Instrumentation amplifiers developments", *AACD workshop*, Pavia, Apr. 9, 2008.

Layout Issues

A

A.1 Introduction

A book describing dynamic offset compensated amplifiers, without information about layout issues would be unfinished. Analog engineers sometimes overlook critical layout issues, therefore, redesigns can become necessary in practical implementations of auto-zero and chopper amplifiers. It is always advisable to do a parasitic extraction in dynamic offset compensated systems.

Contrary to normal amplifiers, dynamic offset compensated amplifiers require on-chip oscillators and digital circuits to generate clocks, which drive the dynamic offset compensation circuits. The clock generation, clock distribution, and switching can severely cripple the offset performance of these amplifiers.

In order to improve common-mode behaviour it is also beneficial to short the backgate of input switches to the sources, as shown in figure A-1. Because the backgate source voltage changes with the input common-mode in figure A-1a, the threshold voltage is effected. This means that the channel charge injection is modulated by the common-mode input voltage, which translates into a CMRR limit. The implementation shown in figure A-1b also has an extended common-mode range.

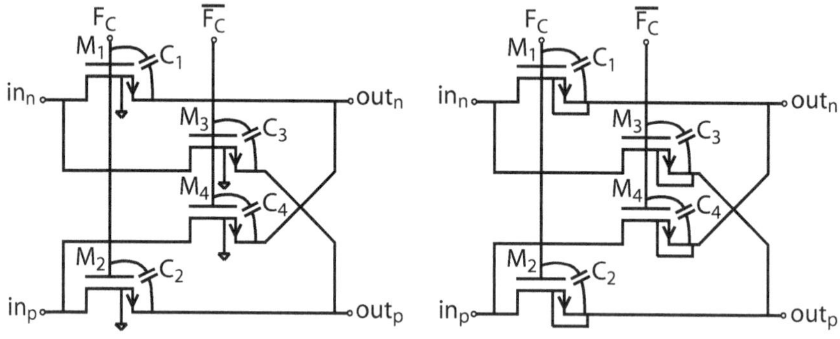

a *b*

Fig. A-1 *Two possible implementations of NMOS choppers (a) with backgates connected to ground, (b) with backgates connected to the sources.*

In section A.2 methodologies are shown to minimize chopper parasitic capacitance mismatches, and section A.3 deals with shielding techniques for clock lines and ground lines. This chapter ends with conclusions in section A.4.

Clock feedthrough

In section 2.3.3 it was illustrated that the mismatch of clock feedthrough capacitances in a differential chopper amplifier can cause a residual offset voltage [A.2]. In section 2.3.3 it was also shown that the residual offset can be expressed by:

$$V_{os,res1} = 2(\mathrm{R}_1+\mathrm{R}_2)(\mathrm{C}_1-\mathrm{C}_2)V_F\mathrm{F}_\mathrm{C}, \quad \text{(A-1)}$$

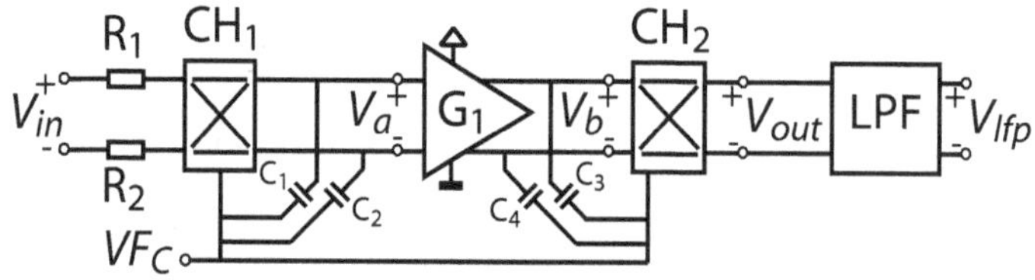

Fig. A-2 *Charge injection model in a chopper amplifier.*

where (R_1+R_2) is the input impedance including ON-resistances of the chopper switches, (C_1-C_2) is the mismatch of clock feed through capacitances, V_F is the driving voltage, and F_C is the chopper frequency. For example a 20 kHz chopper frequency, ON-resistance of 5 kΩ, no source impedance, and a 5V driving voltage would lead to a residual offset per unit of capacitance mismatch between C_1 and C_2 of 2 μV/fF.

Also the capacitors C_3 and C_4 depicted in figure A-2 can cause residual offset. This leads to a residual offset which can be expressed by:

$$V_{os,res2} = \frac{2(C_3 - C_4) V_F F_C}{G_1}, \qquad \text{(A-2)}$$

where G_1 is the transconductance of the chopped amplifier. It can be seen that higher transconductance amplifiers are less vulnerable to the mismatch of C_3 and C_4. For example for a 20 kHz chopper frequency, a 100 μA/V transconductance, and a 5 V driving voltage would also lead to a residual offset per unit of capacitance of 2 μV/fF.

It can be concluded that the mismatch of C_1 and C_2 as well as the mismatch of C_3 and C_4 are both of importance and effort must be made to minimize those mismatches. Therefore, the next section focuses on the layout of choppers.

A.2 Chopper layout

Two layouts of choppers have been developed during the design of the amplifiers discussed in chapters 5 and 6. The analysis in the previous section showed that charge injection mismatch from the input chopper to the input of the amplifier that is chopped and from the output chopper to the output of the amplifier that is chopped is what mainly causes the residual offset.

This charge injection can be reduced by reducing the absolute capacitive crosstalk from clock lines towards the one side of the chopper. In the layout shown in figure A-3 the poly clock lines only run parallel to the input lines, and do not run parallel to the output lines. Except for the parasitic capacitance mismatch due to metal differences, there can also be a mismatch in the gate-source and gate-drain capacitance due to transistor mismatch. To minimize transistor mismatch the use of dummy transistors is widely

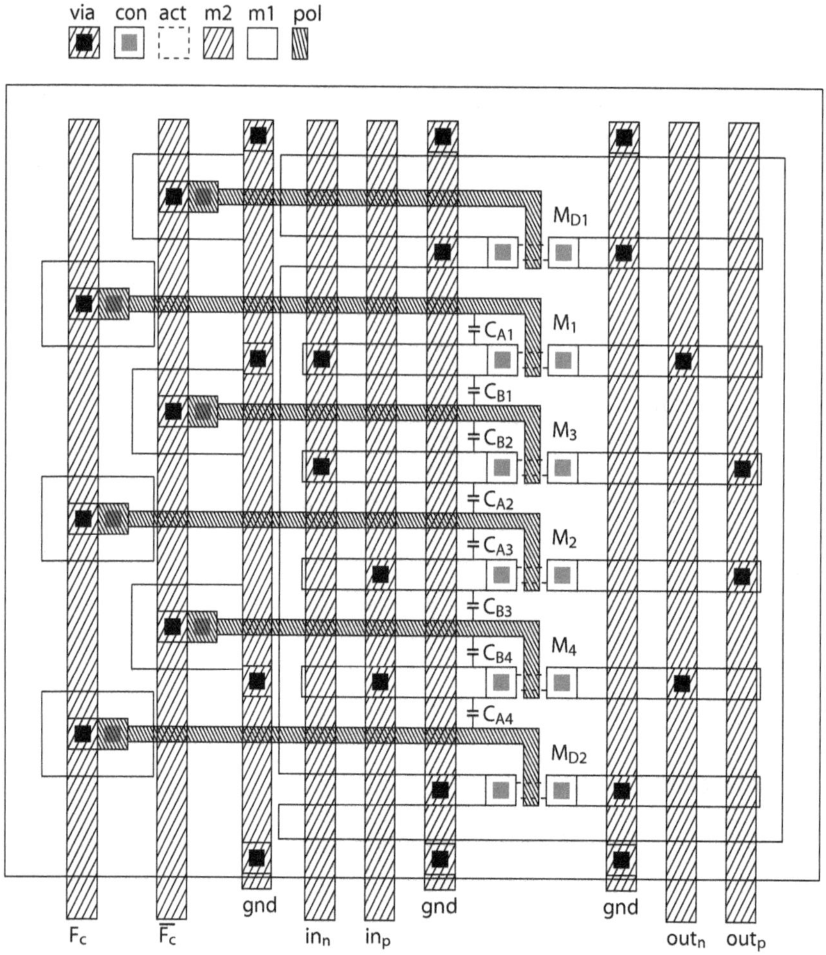

Fig. A-3 *Chopper layout with dummy switches to balance the parasitic capacitances and improve the mismatch of the chopper switches.*

accepted [A.1]. The chopper layout shown in figure A-3 contains two dummy transistors: M_{D1} and M_{D2}. These dummy transistors give transistors M_1–M_4 topologically and electrically the same surroundings. The gate line driving the dummy switch M_{D2} is necessary to make the crosstalk from F_c to in_n equal to the crosstalk of F_c to in_p by making the capacitances $C_{A1}+C_{A2}$ equal to $C_{A3}+C_{A4}$. The gate line driving M_{D1} is not really necessary, although it does help making the parasitic crosstalk from F_c to in_n and in_p and from $\overline{F}_c$ to in_n and in_p equal to each other, which helps in reducing common-mode spikes. To further reduce capacitive clock feedthrough, a grounded metal 1 plate is used to shield the metal 2 clock lines from the signal lines.

In conclusion, the layout depicted in figure A-3 has matched switch transistors, minimised capacitive crosstalk towards the output, and balanced crosstalk from each clockline to the differential inputs and outputs. A disadvantage of this layout is that it needs a large area. By using minimum size switches, the matching will never be very good and so the use of dummy transistors may be regarded as being superfluous.

Another more compact layout was developed without the use of dummy transistors. This layout was made as symmetrically as possible to balance the parasitic capacitances. This layout is shown in figure A-4. In comparison to the previous layout, the switches are closer to each other, which improves transistor mismatch.

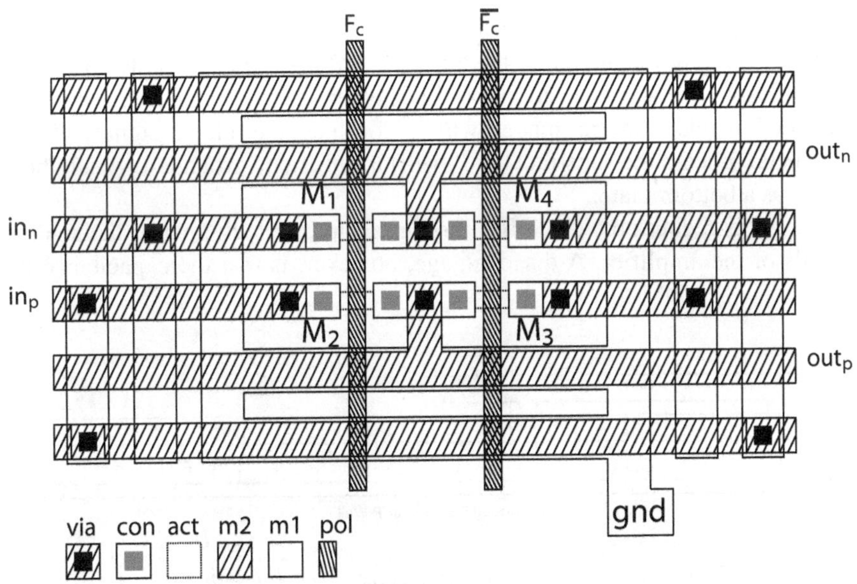

Fig. A-4 *Compact chopper layout with balanced and shielded clock feedthrough capacitances.*

The absolute parasitic capacitance is reduced with respect to the previous layout because, firstly, the signal lines and clock lines are always perpendicular to each other, whereas in figure A-3 they also run parallel to each other. Secondly, it is possible to shield the metal 2 signal lines from the poly clock lines with a grounded metal 1 shield.

During the research described in this book the two layouts were both used, but they were never used in the same amplifier. Therefore, it could not be verified experimentally which layout would benefit residual offset the most. The layout shown in figure A-3 has been used in the operational amplifier design described in section 5.2, all other implementations use the layout depicted in figure A-4.

A.3 Clock shielding

In the previous section, layouts of choppers were discussed. The clock signals have to be routed from the clock generation circuit to the switches of the dynamic offset compensation. This could lead to unforeseen clock feedthrough and substrate feedthrough.

To minimize the clock feedthrough, an on-chip coax cable was used in all designs. A cross section of this layout is shown in figure A-5. A grounded shield can be made, which confines the clock signals. In a two-metal process the substrate (or P-well) itself can be grounded, which then serves as a bottom plate.

This reduces the clock feedthrough from the clock lines to the analog signals of the amplifier. A disadvantage, however, is that the capacitance of

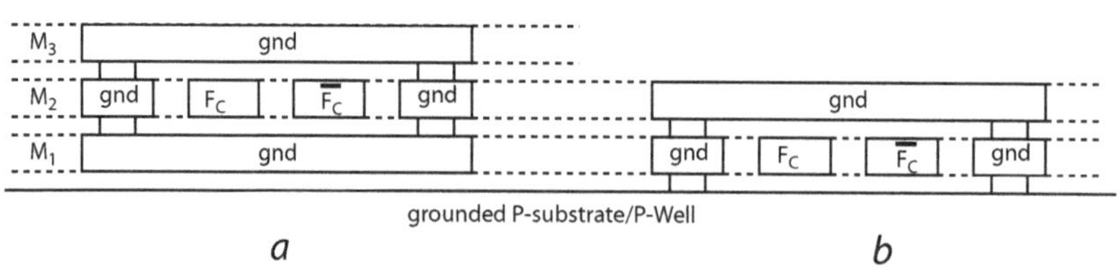

Fig. A-5 *Cross sections of on-chip coax clock lines in (a) a 3-metal process and (b) a 2-metal process.*

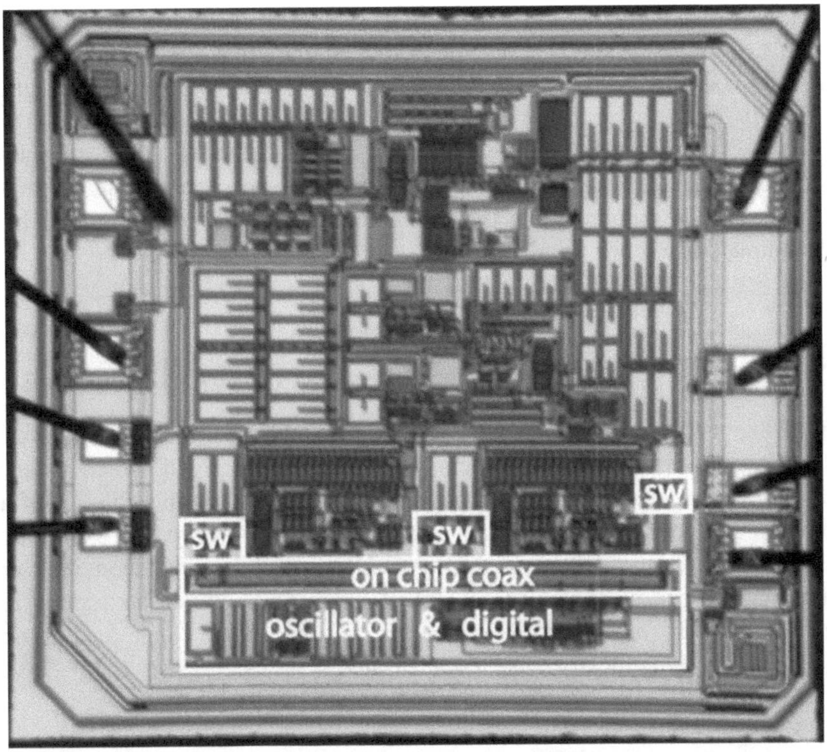

Fig. A-6 *Chip micrograph of the operational amplifier discussed in chapter 5.3 with the oscillator, on-chip coax, and chopper and auto-zero switches, SW, indicated.*

the clock lines to ground increases and relatively strong digital buffers are needed to boost the clock signals.

Another disadvantage is that by using all metal layers available in a process, there are no remaining metal layers to route signals from one side of the shield to the other side of the shield. However, when the layout of a dynamic offset compensated amplifier is planned carefully, crossings are not necessary. When a crossing is unavoidable, the crossing should be created in such a way that the parasitic capacitive feedthrough is balanced. It is also important that the clock generation circuits do not interfere too much with the amplifier. The separation of an analog and digital ground can be advised.

An example of a chip is shown in figure A-6. In this layout the oscillator and digital circuit that generate the clock signals are placed in the bottom. The on-chip coax is used both to shield the clock generation circuits from the amplifier and to shield the clock signals that are confined by the on-chip coax. In this chip the chopper and auto-zero switches are designed similarly to the chopper shown in figure A-4.

In these layouts the ground was separated in four grounds: an analog ground to bias the amplifier, a digital ground to bias the clock generation circuit, a shield ground to bias the on-chip coax, and a separate ground to bias the output stage of the amplifier.

A.4 Conclusion

In this appendix a brief overview was given to address critical layout issues. Two layouts of chopper amplifiers were discussed both of which balance parasitic capacitive clock feedthrough. It was also shown that the use of an on-chip coax makes the shielding of clock lines possible. As an example, a strategy for the layout of one amplifier was discussed. In conclusion, it is very important to carefully plan the layout of a dynamic compensated amplifier, and to be very aware of the problems of unbalanced clock feedthrough capacitors in dynamically offset compensated systems.

A.5 References

[A.1] A. Hastings, "*The art of analog layout*", New Jersey: Prentice Hall, 2001.

[A.2] A. Bakker, J.H. Huijsing. "*High-accuracy CMOS smart temperature sensors*", Dordrecht: Kluwer, 2000.

About the Authors

J.F. (Frerik) Witte was born in Amsterdam, the Netherlands, on March 16, 1979, where he lived until finishing his high school education (Atheneum) at the Pieter Nieuwland College in 1997. In that year he moved to Delft to start his studies in electrical engineering. He received his M.Sc. degree in electrical engineering (cum laude) from Delft University of Technology in 2003. The subject of his M.Sc. thesis was "On-Chip Time References and Electro-Thermal Oscillators". In 2003, he started working towards a Ph.D. degree at the Electronic Instrumentation Laboratory of Delft University of Technology. The subject of his research was to design low-offset broadband CMOS amplifiers, which resulted in this book.

From January to April 2003, he did an internship at Philips Semiconductors, San Jose, California, where he worked on integrated diagnostic circuits to measure the value of an external capacitor. Since January 2009, he is working as a senior design engineer for National Semiconductor at Delft. His professional interests include sensors, precision analog and mixed-signal design. He received the ESSCIRC 2006 Young Scientist Award.

Kofi A. A. Makinwa received the B.Sc. and M.Sc. degrees from Obafemi Awolowo University, Ile-Ife, Nigeria, in 1985 and 1988, respectively, the M.E.E. degree from the Philips International Institute, Eindhoven, the Netherlands, in 1989, and the Ph.D. degree from Delft University of Technology, Delft, the Netherlands, in 2004.

From 1989 to 1999, he was a Research Scientist with Philips Research Laboratories, where he designed sensor systems for interactive displays and analog front-ends for optical and magnetic recording systems. In 1999, he joined Delft University of Technology, where he is currently an professor with the Faculty of Electrical Engineering, Computer Science and Mathematics. His main research interests are in the design of precision analog circuitry, sigma-delta modulators and sensor interfaces. His work has resulted in ten U.S. patents and over 70 technical papers.

Prof. Makinwa is on the program committees of several international conferences, including the IEEE International Solid-State Circuits Conference (ISSCC) and the International Solid-state Sensors and Actuators Conference (Transducers). He has given plenary talks and tutorials at several conferences, including twice at the ISSCC. He is a co-recipient of JSSC (2005), ISSCC (2008, 2006, 2005), ESSCIRC (2006) and ISCAS (2008) best paper awards. In 2005, he received the Veni Award from the Netherlands Organization for Scientific Research and the Simon Stevin Gezel Award from the Dutch Technology Foundation. He is a distinguished lecturer of the IEEE Solid-State Circuits Society and a fellow of the Young Academy of the Royal Netherlands Academy of Arts and Sciences.

Johan H. Huijsing was born on May 21, 1938. He received the M.Sc. degree in Electrical Engineering from the Delft University of Technology, Delft, the Netherlands in 1969, and the Ph.D. degree from this University in 1981 for his thesis on operational amplifiers.

He has been an assistant and associate professor in Electronic Instrumentation at the Faculty of Electrical Engineering of the Delft University of Technology since 1969, where he became a full professor in the chair of Electronic Instrumentation since 1990, and professor-emeritus since 2003. From 1982 through 1983 he was a senior scientist at Philips Research Labs. in Sunnyvale, California, USA. From 1983 until 2005 he was a consultant for Philips Semiconductors, Sunnyvale, California, USA and since 1998 also a consultant for Maxim, Sunnyvale, California, USA.

The research work of Johan Huijsing is focused on the systematic analysis and design of operational amplifiers, analog-to-digital converters and integrated smart sensors. He is author or co-author of some 250 scientific papers, 40 patents and 13 books, and co-editor of 13 books. As a professor he guided 27 Ph.D. students toward their degree. He is fellow of IEEE for contributions to the design and analysis of analog integrated circuits. He was awarded the title of Simon Stevin Meester for applied Research by the Dutch Technology Foundation.

Index

A

auto-zero 14–23
  auxiliary input stage 17
  input offset storage 16
  instrumentation amplifier 76
  noise 19
  output offset storage 15

B

biasing
  constant current 141
  constant transconductance 95
  PTAT 140
  PTIM 95

C

charge injection 15, 17, 18, 27, 29, 31, 32, 33, 34, 35, 36, 37, 40
  see also residual offset
chopped auto-zero amplifier 29–31
chopper 23–29
  amplifier 24
  charge injection 27–29
  feedback amplifier 26
  instrumentation amplifier 75
  noise 25
  ripple 88
  ripple reduction 88
chopper layout 158, 159
chopper offset-stabilized chopper amplifier 60, 82
common-mode feedback
  input controlled 111
  long tailed pair 92, 109, 110
  resistive 94, 142
continuous-time auto-zero
  see also offset-stabilization auto-zero 48
cross talk capacitors in choppers 27–29
current-feedback instrumentation amplifier 68–74
  common-mode gain dependency 129, 146
current-sensing 129–133

D

dynamic offset compensation
  see auto-zero
  see chopper
  see offset-stabilization

F

frequency compensation
  Miller 94
  multi-path hybrid-double-nested Miller 52, 53
  multi-path hybrid-nested

Miller 50, 51, 54, 55, 56, 57, 61, 88, 98, 104, 138
multi-path nested Miller 50
nested Miller 55

G

gain-boosting 110

H

high-side current-sensing 130

L

low-side current-sensing 130

M

mismatch 4–7

N

noise
effect of auto-zero on 19–23
effect of chopper and auto-zero offset-stabilization on 59
effect of chopper offset stabilization on 49
effect of chopper on 25
noise folding 19–23

O

offset 3
minimizing 6
offset in CMOS amplifiers 3
offset-stabilization 45–59
auto-zero 47
chopper 48
chopper and auto-zero 58
chopper with ripple filter 55–58
concept 45
frequency compensation 50
instrumentation amplifier 79–81
iterative 61
on-chip coax 160
opamp
class-A 94, 142
class-AB 92, 108
folded cascode 6, 92, 109
gain-boosted folded cascode 110

P

ping-pong 44–45, 78
ping-pong-pang 78–79
PTAT current source 140
PTIM current source 95

R

residual offset
due to charge injection 15, 17, 18, 27, 88
due to chopped parasitic 54, 56, 89, 106, 120
due to finite gain 17, 18, 46, 47, 56, 57, 87
due to parasitics 28, 29, 156, 157
layout 156–162
ripple reduction 88

T

threshold voltage mismatch 4, 5
transconductance factor mismatch 4, 5

FSC
www.fsc.org
MIX
Papier | Fördert gute Waldnutzung
FSC® C083411

Kolibri 360 GmbH
Ferdinand-Jühlke-Straße 7
99095 Erfurt, Deutschland
produktsicherheit@kolibri360.de